KNOW YOUR BODY

The Essential Guide to Human Anatomy and Physiology

WORKBOOK

MARY DALGLEISH

PAGE TURNER AWARDS
Writing Award Finalist 2020

PAGE TURNER AWARDS
FINALIST 2021

First Printing, November 2020

ISBN: 978-1-8380641-0-5

https://www.anatomyandphysiology.co.uk

Disclaimer

This book is for educational purposes only. It is not meant as medical advice, diagnosis or treatment and you must not rely on the information in this book as an alternative or substitute to medical advice from your doctor or other professional healthcare provider.

If you have specific questions about any medical matter, you should consult your doctor or other professional healthcare provider.

If you think you may be suffering from any medical condition, you should seek immediate attention from your doctor or other professional healthcare provider.

You should never delay in seeking medical advice, disregard medical advice or discontinue medical treatment based on information in this book.

To the extent that you rely on any information contained in this book, you assume all risks involved in such reliance.

NOTE: All website links were checked and active in October 2020. However, due to the dynamic nature of the Internet, any web addresses or links contained in this book may have changed since publication, and may no longer be valid. Please exercise due diligence in regard to any website content.

UK spelling is used throughout this text and every effort has been made to ensure accuracy.

Introduction

Welcome to the "KNOW YOUR BODY The Essential Guide to Human Anatomy and Physiology" WORKBOOK. It is designed to check your knowledge of this subject and to help you revise for your exams. You can use it alongside the companion book or in conjunction with any other anatomy and physiology textbook.

You will find multiple choice tests on various topics including each body system, as well as mixed multiple choice tests of 50 questions. There are also word searches, gap filling exercises and crosswords.

Answers to all these exercises are provided at the back of the book along with study tips and a list of web resources.

Contents

Anatomical Terminology – circle correct answer

1. The study of the structure and relationship between body parts is termed:
   a) Pathology
   b) Physiology
   c) Anatomy
   d) Pathophysiology

2. The study of disease is called:
   a) Pathology
   b) Anatomy
   c) Physiology
   d) Homeostasis

3. The state of internal balance in the body is called:
   a) Pathology
   b) Metabolism
   c) Physiology
   d) Homeostasis

4. The suffix –**itis** as in arthr**itis** means:
   a) Disease
   b) Inflammation
   c) Pain
   d) Heat

5. Medial means:
   a) Closer to the midline of the body
   b) Further away from the midline of the body
   c) Close to the head
   d) Further from the trunk

6. The sagittal plane divides the body or an organ into:
   a) Quadrants
   b) Front (anterior) and back (posterior) regions
   c) Right and left regions
   d) Upper (superior) and lower (inferior) regions

7. This is part of the axial skeleton:
   a) Spine
   b) Hand
   c) Foot
   d) Scapula

8. This is part of the appendicular skeleton:
   a) Ribcage
   b) Spine
   c) Leg
   d) Skull

9. The iliac region of the abdomen is:
   a) Near the ribs
   b) Near the hip bones
   c) Around the navel
   d) Near the liver

10. 'Glosso' refers to the:
    a) Brain
    b) Eyes
    c) Tongue
    d) Cartilage

11. This refers to a type of disease that is transferred from animals to humans:
    a) Zoonotic
    b) Bacterial
    c) Viral
    d) Infectious

12. Something that may increase susceptibility to disease is called a:
    a) Risk factor
    b) Prognosis
    c) Red flag sign
    d) Syndrome

13. A disease or disorder of unknown origin is referred to as:
    a) Chronic
    b) Acute
    c) Idiopathic
    d) Communicable

14. Refers to the likely outcome of a disease or ailment:
    a) Risk factor
    b) Result
    c) Prognosis
    d) Syndrome

Body Compounds – circle correct answer

1. These 4 elements make up 96% of human body mass:
   a) Oxygen, Carbon, Hydrogen & Nitrogen
   b) Oxygen, Carbon Dioxide, Proteins & Fats
   c) Protein, Fats, Carbohydrates & Minerals
   d) Water, Vitamins, Minerals & Oxygen

2. Sugars and starches used as fuel by the body are called:
   a) Fats
   b) Carbohydrates
   c) Proteins
   d) Lipids

3. The proteins myosin and actin are found in:
   a) Bone
   b) Fat
   c) Muscle
   d) Blood

4. This element is important for healthy thyroid function:
   a) Iodine
   b) Zinc
   c) Sodium
   d) Chlorine

5. This element is needed for healthy bones and teeth:
   a) Chloride
   b) Iron
   c) Calcium
   d) Iodine

6. This element is important for regulating water balance:
   a) Iodine
   b) Zinc
   c) Sodium
   d) Magnesium

7. This is a critical ingredient in haemoglobin:
   a) Cobalt
   b) Copper
   c) Zinc
   d) Iron

8. The pH scale measures:
   a) Density
   b) Weight
   c) Acidity and alkalinity
   d) Blood count

9. The pH of pure water is:
   a) 5
   b) 6
   c) 7
   d) 10

10. These are fat-soluble vitamins:
    a) A, B, C, D
    b) B1, B2, C, D
    c) A, D, E, K
    d) B & C

11. Vitamin important for blood clotting and often given to babies after birth:
    a) C
    b) K1
    c) B12
    d) D

12. This vitamin is produced in the skin by the action of sunlight:
    a) Vitamin B
    b) Vitamin A
    c) Vitamin B12
    d) Vitamin D3

13. This vitamin is also known as ascorbic acid:
    a) C
    b) K1
    c) B12
    d) D

14. This vitamin is also known as alpha-tocopherol:
    a) Vitamin B
    b) Vitamin A
    c) Vitamin E
    d) Vitamin D3

Histology – circle correct answer

1. The control centre of a cell is called the:
   a) Mitochondria
   b) Nucleus
   c) Nucleolus
   d) Centrosome

2. Pinocytosis means:
   a) Cell drinking – taking in fluid
   b) Cell eating – absorbing solid materials such as bacteria
   c) Movement of molecules across cell membrane
   d) Destruction of harmful organisms

3. Movement of water across cell membranes is referred to as:
   a) Active transport
   b) Osmosis
   c) Dissolution
   d) Diffusion

4. The power stations of a cell that create energy are called:
   a) Centrioles
   b) Lysosomes
   c) Ribosomes
   d) Mitochondria

5. The protein builders of a cell are called:
   a) Centrioles
   b) Lysosomes
   c) Ribosomes
   d) Mitochondria

6. Sometimes referred to as the waste disposal units within cells:
   a) Centrioles
   b) Lysosomes
   c) Ribosomes
   d) Mitochondria

7. Telomeres are:
   a) Stable chromosome ends that protect the DNA
   b) Parts of the chromosomes that get shorter with each cell division
   c) Inherited from our parents
   d) All of the above

8. In prophase of mitosis (cell division):
   a) The cell divides into two separate daughter cells
   b) The replicated chromatin or DNA strands (chromatids) form into 46 X shaped chromosomes
   c) The chromosomes line up at the centre of the cell
   d) The cell is resting

9. The ova and sperm cells that undergo meiosis have:
   a) 46 chromosomes
   b) 21 chromosomes
   c) 92 chromosomes
   d) 23 chromosomes

10. An unusual arrangement of chromosomes is called a:
    a) Cytokinesis
    b) Division
    c) Translocation
    d) Prion

11. This tissue covers body surfaces and lines body cavities:
    a) Connective tissue
    b) Epithelial tissue
    c) Nervous tissue
    d) Muscle tissue

12. This tissue envelops, separates, or binds structures of the body:
    a) Fascia (connective tissue)
    b) Epithelial tissue
    c) Nervous tissue
    d) Muscular tissue

13. Membrane that secretes fluid and found in joints:
    a) Serous
    b) Mucous
    c) Synovial
    d) Periosteum

14. Cells that are the support systems for nerves:
    a) Neurons
    b) Epithelial cells
    c) Adipose cells
    d) Neuroglial cells

Skin – circle correct answer

1. The deepest layer of the epidermis is:
    a) The stratum spinosum or prickle-cell layer
    b) The stratum corneum or horny layer
    c) The stratum germinativum or basal layer
    d) The stratum granulosum or granular layer

2. Collagen and elastin fibres are most abundant in:
    a) The epidermis
    b) The subcutaneous layer of the epidermis
    c) The clear layer of the epidermis
    d) The dermis

3. These cells secrete histamine in allergic reactions:
    a) Mast cells
    b) Phagocytes
    c) Melanocytes
    d) Basal cells

4. 'Protein glycation' is damage to proteins in the dermis caused by:
    a) Excess protein
    b) Excess sugar
    c) Excess fat
    d) Excess water

5. These glands in the armpit become active at puberty:
    a) Eccrine sweat glands
    b) Apocrine sweat glands
    c) Sebaceous glands
    d) Mammary glands

6. Antioxidant vitamin with benefits for skin health and appearance:
    a) Vitamin K1
    b) Vitamin B6
    c) Vitamin E
    d) Vitamin B12

7. Loss of melanocytes in areas of the skin results in:
    a) Chloasma
    b) Vitiligo
    c) Alopecia
    d) Rosacea

8. An autoimmune skin disease caused by an overproduction of collagen:
    a) Lupus
    b) Eczema
    c) Scleroderma
    d) Dermatitis

9. This skin condition resulting from increased epidermal cell turnover:
    a) Lupus
    b) Eczema
    c) Scleroderma
    d) Psoriasis

10. This virus causes chicken pox:
    a) Varicella Zoster
    b) Herpes Simplex
    c) Coronavirus
    d) Herpes Zoster

11. Tinea pedis is:
    a) Body lice
    b) Head Lice
    c) Athlete's foot
    d) Body ringworm

12. The most serious type of skin cancer is:
    a) Basal Cell carcinoma
    b) Squamous cell carcinoma
    c) Rodent ulcer
    d) Malignant melanoma

13. Hemostasis means:
    a) Excess flow of blood
    b) A haemorrhage
    c) Stopping a flow of blood
    d) Inflammation of the skin

14. Deep bacterial skin infections that start in hair follicles:
    a) Boils
    b) Acne
    c) Rosacea
    d) Eczema

Hair – circle correct answer

1. The cells containing keratin and melanin are found in this part of the hair:
   a) The medulla
   b) The cortex
   c) The root
   d) The cuticle

2. The 'active' or growth phase of hair is called:
   a) Catagen
   b) Telogen
   c) Anagen
   d) Interphase

3. The fine, soft, downy hair of the foetus is called:
   a) Lanugo
   b) Peace fuzz hair
   c) Terminal hair
   d) Telogen hair

4. Loss of patches of scalp hair is called:
   a) Alopecia totalis
   b) Alopecia areata
   c) Alopecia universalis
   d) Traction alopecia

5. Excessive hair growth in women is called:
   a) Folliculitis
   b) Alopecia
   c) Hirsutism
   d) Effluvium

6. Male pattern baldness is called:
   a) Traction alopecia
   b) Alopecia areata
   c) Androgenic alopecia
   d) Alopecia totalis

7. Inflammation of the hair follicles is:
   a) Impetigo
   b) Folliculitis
   c) Alopecia
   d) Rosacea

8. Hair loss can be caused by:
    a) Chemotherapy
    b) Shock and stress
    c) Nutritional deficiencies
    d) All of the above

9. Complete hair loss, including eyebrows, eyelashes and body hair is called:
    a) Traction alopecia
    b) Alopecia areata
    c) Androgenic alopecia
    d) Alopecia totalis

10. Sycosis barbae is:
    a) Inflammation of hair follicles in the beard area
    b) Hair loss in the beard area
    c) Shaving rash
    d) Itching in the beard area

11. Pediculosis capitis is another name for:
    a) Scabies
    b) Body lice
    c) Pubic lice
    d) Head lice

12. This condition can cause hirsutism in females:
    a) Psoriasis
    b) Polycystic ovarian syndrome
    c) Head lice
    d) Body ringworm

13. Glands attached to hair follicles that produce natural oils:
    a) Sweat glands
    b) Salivary glands
    c) Sebaceous glands
    d) Cowper's glands

14. Resting phase of hair growth where no active growth occurs:
    a) Anagen
    b) Catagen
    c) Telogen
    d) Interphase

Nails – circle correct answer

1. The soft, pink tissue that sits underneath the nail is called the:
   a) Nail bed
   b) Nail plate
   c) Free edge of nail
   d) Hyponychium

2. This part of the nail is commonly known as the cuticle:
   a) Hyponychium
   b) Paronychium
   c) Eponychium
   d) Lunula

3. Each month, fingernails grow at an average rate of:
   a) 10 millimetres
   b) 1 millimetre
   c) 3 millimetres
   d) 7 millimetres

4. The technical or medical term for a nail disease is:
   a) Onychauxis
   b) Onycholysis
   c) Onychosis
   d) Onychocryptosis

5. The technical term for a spoon shaped or concave nail is:
   a) Koilonychia
   b) Onycholysis
   c) Onychosis
   d) Onychocryptosis

6. The technical or medical term for an ingrown toenail is:
   a) Onychauxis
   b) Onycholysis
   c) Onychosis
   d) Onychocryptosis

7. White spots or streaks on nails are called:
   a) Cyanosis
   b) Lunula
   c) Leukonychia
   d) Paronychia

8. Tinea unguium is:
   a) Hang nail
   b) Inflammation of the nail
   c) Ringworm of the nail (fungal infection)
   d) Deep ridges on the nail

9. Bacterial infection of the skin surrounding the nail is called:
   a) Onychia
   b) Lunula
   c) Leukonychia
   d) Paronychia

10. Onychorrhexis refers to:
    a) Hangnails
    b) Ingrown nails
    c) Nail pitting
    d) Dry, brittle, splitting nails

11. The technical term for nail biting is:
    a) Koilonychia
    b) Onycholysis
    c) Onychophagia
    d) Onychocryptosis

12. Cyanosis is when the fingernails appear:
    a) White
    b) Red
    c) Blue
    d) Green

13. This skin condition commonly affects the nails:
    a) Acne
    b) Eczema
    c) Chloasma
    d) Psoriasis

14. Whitlow - a painful and highly contagious infection is caused by:
    a) A bacterial infection
    b) A herpes simplex viral infection
    c) A fungal infection
    d) A parasitic infection

Skeletal System – circle correct answer

1. How many bones in an adult skeleton?
   a) 106
   b) 206
   c) 220
   d) 198

2. This is part of the appendicular skeleton:
   a) Ribcage
   b) Spine
   c) Shoulder girdle (scapula and clavicles)
   d) Skull

3. These cells break down bone tissue:
   a) Osteoblasts
   b) Osteocytes
   c) Osteoclasts
   d) Osteoporosis

4. This is a fracture where a bone is broken in several places:
   a) Simple fracture
   b) Impacted fracture
   c) Communited fracture
   d) Spiral fracture

5. Loss of bone density leads to a condition called:
   a) Osteoarthritis
   b) Osteogenesis imperfecta
   c) Osteomalacia
   d) Osteoporosis

6. This disease results in enlarged and misshapen bones:
   a) Osteogenesis imperfecta
   b) Paget's disease
   c) Osteomalacia
   d) Rickets

7. This smooth cartilage covers the ends of bones at a joint:
   a) Fibrocartilage
   b) Hyaline cartilage
   c) Yellow elastic cartilage
   d) Costal cartilage

8. The joints between the carpal bones and tarsal bones are:
   a) Hinge joints
   b) Saddle joints
   c) Plane or gliding joints
   d) Condyloid joints

9. This is regarded as an autoimmune disease:
   a) Rheumatoid arthritis
   b) Osteoarthritis
   c) Osteomalacia
   d) Paget's disease

10. A butterfly-shaped bone forming the floor of the cranium and the temples:
    a) Ethmoid
    b) Mandible
    c) Maxilla
    d) Sphenoid

11. This is a sideways curvature of the spine:
    a) Kyphosis
    b) Lordosis
    c) Scoliosis
    d) Spinal stenosis

12. This is an inward curvature of the lumbar spine:
    a) Kyphosis
    b) Lordosis
    c) Scoliosis
    d) Spinal stenosis

13. This condition causes severe pain in the heel area:
    a) Plantar fasciitis
    b) Morton's neuroma
    c) Hammertoe
    d) Prepatellar bursitis

14. This condition is caused by irritation to a nerve in the foot:
    a) Plantar fasciitis
    b) Morton's neuroma
    c) Hammertoe
    d) Sciatica

Muscular System – circle correct answer

1. Voluntary or striated/striped muscle tissue is also known as:
   a) Cardiac muscle
   b) Skeletal muscle
   c) Smooth muscle
   d) Visceral muscle

2. How many types of muscle tissue are there in the human body?
   a) Two
   b) Four
   c) Three
   d) Six

3. This connective tissue covers individual muscles:
   a) Epimysium
   b) Perimysium
   c) Endomysium
   d) Periosteum

4. This neurotransmitter is involved in muscle movements:
   a) Serotonin
   b) Noradrenaline
   c) Histamine
   d) Acetylcholine

5. A disorder involving excessive muscle activity:
   a) Dystonia
   b) Fibromyalgia
   c) Muscular dystrophy
   d) Myasthenia gravis

6. Muscular dystrophy is:
   a) A viral disorder
   b) A bacterial disorder
   c) An autoimmune disorder
   d) A genetic disorder

7. A painful knee condition that sometimes affects adolescents:
   a) Adhesive capsulitis
   b) Osgood-Schlatter syndrome
   c) Myasthenia Gravis
   d) Fibromyalgia

8. Actin and Myosin are made of:
   a) Enzymes
   b) Minerals
   c) Vitamins
   d) Proteins

9. Atrophy is a condition in which muscles:
   a) Lack tone
   b) Become inflamed
   c) Waste
   d) Lack sodium

10. Lactic acid build up in muscles can result from:
    a) Excessive exercise without a break
    b) Excess acid in the diet
    c) Lack of sleep
    d) Lack of exercise

11. Fascia is:
    a) Fibrous connective tissue
    b) A covering for muscles
    c) Internal connective tissue that wraps around organs
    d) All of the above

12. Antagonist refers to:
    a) A contracted muscle
    b) A relaxed muscle
    c) An insured muscle
    d) An overworked muscle

13. A condition that causes widespread muscle and joint pain and fatigue:
    a) Adhesive capsulitis
    b) Fibromyalgia
    c) Muscular dystrophy
    d) Myasthenia gravis

14. These muscles are affected by 'tennis elbow':
    a) Forearm flexor muscles
    b) Finger flexor muscles
    c) Forearm extensor muscles
    d) Shoulder muscles

Skeletal Muscles – circle correct answer

1. An action of the gluteal muscles is to:
   a) Move the thigh
   b) Flex the knee
   c) Flex the hip
   d) Plantar flex the foot

2. The muscles of the lower arm include:
   a) Tibialis anterior
   b) Peroneus longus
   c) Extensor carpi digitorum
   d) Soleus

3. The action of the masseter muscle is to:
   a) Raise the lower jaw and clench the teeth
   b) Lift the upper lip
   c) Flex the neck
   d) Compress the nostrils

4. What is the action of the rhomboids muscles?
   a) Stabilise the humerus
   b) Adduct the scapulae
   c) Extend the vertebral column
   d) Support the viscera

5. In which muscle's tendon of insertion is the patella found?
   a) Biceps femoris
   b) Biceps brachii
   c) Gastrocnemius
   d) Quadriceps femoris

6. The quadriceps femoris muscles:
   a) Flex the knee
   b) Abduct the leg
   c) Extend the knee
   d) Adduct the leg

7. Semimembranosus, Semitendinosus & Biceps Femoris are:
   a) The hamstring muscles
   b) The quadriceps muscles
   c) The adductor muscles
   d) The abductor muscles

8. Which muscle flexes the big toe?
   a) Tibialis posterior
   b) Flexor hallucis longus
   c) Tibialis anterior
   d) Peroneus longus

9. What is the main extensor of the elbow?
   a) Biceps femoris
   b) Biceps brachii
   c) Triceps brachii
   d) Trapezius

10. A muscle that decreases the angle between two bones is known as:
    a) An adductor
    b) An abductor
    c) An extensor
    d) A flexor

11. This muscle is involved in smiling:
    a) Depressor anguli oris
    b) Mentalis
    c) Zygomaticus major
    d) Temporalis

12. This muscle flexes the neck and turns the head:
    a) Rhomboids
    b) Supraspinatus
    c) Infraspinatus
    d) Sternocleidomastoid

13. The piriformis is a small muscle located deep in the:
    a) Upper arm
    b) Forearm
    c) Buttock
    d) Thigh

14. This is the deepest abdominal muscle located in the lower back:
    a) Rhomboids
    b) Rectus abdominis
    c) Quadratus lumborum
    d) Sternocleidomastoid

Cardiovascular System – circle correct answer

1. The function of venules is to:
   a) Carry oxygenated blood from capillaries to larger veins
   b) Carry oxygenated blood from larger veins to capillaries
   c) Carry deoxygenated blood from capillaries to larger veins
   d) Carry deoxygenated blood from larger veins to capillaries

2. The function of thrombocytes is to:
   a) Produce antibodies
   b) Protect the body from infection
   c) Eat bacteria & other microorganisms
   d) Clot the blood

3. Erythrocytes are:
   a) Cells that fight infection
   b) Cells that are responsible for blood clotting
   c) Cells that transport oxygen
   d) Cells that form collagen

4. Leucocytes are:
   a) Cells that fight infection
   b) Cells that are responsible for blood clotting
   c) Cells that transport oxygen
   d) Cells that form collagen

5. The three layers of the heart are:
   a) Endomysium, epimysium, perimysium
   b) Dura mater, arachnoid mater, pia mater
   c) Endocardium, myocardium, pericardium
   d) Endosteum, episteum, periosteum

6. These blood vessels carry oxygenated blood from the lungs to the heart:
   a) Pulmonary arteries
   b) Pulmonary veins
   c) Aorta
   d) Superior vena cava

7. Where in the heart does the inferior vena cava bring its blood?
   a) Right atrium
   b) Right ventricle
   c) Left atrium
   d) Left ventricle

8. Capillaries:
   a) Arise from arterioles and flow into venules
   b) Arise from arteries and flow into veins
   c) Arise from venules and flow into arterioles
   d) Arise from veins and flow into arteries

9. Which blood group is the universal donor and can give blood to anyone?
   a) B positive +
   b) AB negative -
   c) O negative -
   d) O positive +

10. A person with this type of blood can receive any type of blood:
    a) B positive +
    b) AB negative -
    c) O negative -
    d) O positive +

11. A degenerative disease causing hardening of the arteries is called:
    a) Phlebitis
    b) Atherosclerosis
    c) Arteriosclerosis
    d) Aneurysm

12. A condition that affects the blood's ability to clot normally is called:
    a) Hepatitis
    b) Hemophilia
    c) Hypotension
    d) Hemorrhoids

13. Build-up of cholesterol and fatty deposits inside arterial walls:
    a) Phlebitis
    b) Atherosclerosis
    c) Arteriosclerosis
    d) Aneurysm

14. Inflammation of the veins:
    a) Phlebitis
    b) Varicose veins
    c) Arteriosclerosis
    d) Atherosclerosis

Lymphatic System – circle correct answer

1. Lymph is emptied back into the bloodstream through the
   a) Lymphatic capillaries
   b) Spleen
   c) Lymph nodes
   d) Lymphatic ducts

2. The cervical lymph nodes are situated:
   a) In the front of the ear
   b) Behind the ear
   c) Under the chin
   d) On the neck

3. Peyer's Patches is lymphatic tissue located in the:
   a) Stomach
   b) Large intestine
   c) Small intestine
   d) Armpits

4. The popliteal lymph nodes are located:
   a) Behind the knees
   b) Under the chin
   c) Under the armpits
   d) In the elbow creases

5. Where does lymph from right leg eventually drain?
   a) Right lymphatic duct
   b) Thoracic duct
   c) Cervical lymph nodes
   d) Submandibular lymph nodes

6. The axillary lymph nodes are located:
   a) Behind the knees
   b) Behind the ears
   c) Under the arms
   d) In the groin

7. Lymph flow is hindered or slowed down by:
   a) Muscular activity
   b) Infection
   c) Arterial pulsation
   d) Peristaltic action

8. The role of B lymphocytes is to:
   a) Produce antibodies
   b) Attack antibodies
   c) Destroy antigens
   d) Produce antigens

9. The role of T lymphocytes is to:
   a) Produce antibodies
   b) Actively destroy infected cells
   c) Attack antibodies
   d) Produce antigens

10. Chyle is made up of:
   a) Lymph and blood
   b) Lymph and antibodies
   c) Lymph and digested fat
   d) Lymph and leucocytes

11. This is a cancer of lymphatic tissue:
   a) Grave's disease
   b) Hodgkin's disease
   c) Addison's disease
   d) Cushing's syndrome

12. Swelling caused by compromised lymphatic system and lymph build up:
   a) Lyme Disease
   b) Hemophilia
   c) Lymphoma
   d) Lymphoedema (Lymphedema)

13. Disease caused by the 'Epstein Barr' virus:
   a) Grave's disease
   b) Hodgkin's disease
   c) Glandular fever
   d) Cushing's syndrome

14. Disease affecting lymph nodes and arteries in children under 5:
   a) Grave's disease
   b) Hodgkin's disease
   c) Addison's disease
   d) Kawasaki disease

Immunity – circle correct answer

1. Present from birth and does not have an immunological memory:
   a) Adaptive immunity
   b) Autoimmunity
   c) Innate immunity
   d) Acquired immunity

2. The body's first line of defense against pathogens:
   a) White blood cells
   b) Inflammation
   c) The skin and mucous membranes
   d) On the neck

3. These cells are involved in allergic reactions:
   a) B-Cells
   b) T-Cells
   c) Thrombocytes
   d) Mast cell and basophils

4. These cells help combat parasites in the body:
   a) Basophils
   b) Monocytes
   c) Eosinophils
   d) Dendritic cells

5. Send signals to other cells to heighten their anti-viral defenses:
   a) Interferons
   b) Cytokines
   c) Fibrinogen
   d) Antigens

6. Substances that cause an immune response in the body:
   a) Interferons
   b) Cytokines
   c) Fibrinogen
   d) Antigens

7. This part of the immune system has a great memory:
   a) Adaptive or acquired immunity
   b) Autoimmunity
   c) Innate immunity
   d) Birth immunity

8. Substances produced by B-Cells that attach to the surface of foreign agents in the body and attract other defensive molecules:
   a) Antibodies
   b) Interferons
   c) Cytokines
   d) Antigens

9. These cells mature and develop in the thymus gland:
   a) B-Cells
   b) T-Cells
   c) Thrombocytes
   d) Mast cell and basophils

10. Disease causing increased numbers of immature or abnormal leucocytes:
   a) Lymphoedema
   b) Lymphoma
   c) Leukemia
   d) Lyme Disease

11. Disease that destroys the body's CD4 helper T-cells:
   a) Acquired immune deficiency syndrome (AIDS)
   b) Lymphoma
   c) Leukemia
   d) Glandular fever

12. Study of links between the nervous and the immune systems:
   a) Immunotherapy
   b) Psychoneuroimmunology
   c) Neurology
   d) Immunology

13. Human Immunodeficiency Virus (HIV) is spread through:
   a) Contaminated food
   b) Body fluids
   c) Touch
   d) Inhaled droplets

14. Most common type of white blood cells found in the body:
   a) Monocytes
   b) Basophils
   c) Neutrophils
   d) Eosinophils

Respiratory System – circle correct answer

1. Which of these organs serves both the digestive & respiratory systems?
   a) Trachea
   b) Pharynx
   c) Larynx
   d) Oesophagus

2. The vocal cords are situated in the:
   a) Pharynx
   b) Bronchi
   c) Trachea
   d) Larynx

3. The exchange of gases takes place in the:
   a) Bronchioles
   b) Bronchi
   c) Pleura
   d) Alveoli

4. The structure of the pleura is:
   a) Yellow elastic connective tissue
   b) Areolar tissue
   c) Serous membrane
   d) Columnar epithelial tissue

5. Part of the brain involved with respiration:
   a) Cerebellum
   b) Cerebrum
   c) Medulla oblongata
   d) Pineal gland

6. The respiratory centre is normally stimulated by:
   a) Nitrogen in the blood
   b) Blood pressure
   c) Actions of the diaphragm
   d) Carbon dioxide in the blood

7. This nerve controls the diaphragm:
   a) Vagus
   b) Olfactory
   c) Phrenic
   d) Intercostal

8. Which of the following is NOT in the thoracic cavity?
   a) Heart
   b) Common iliac artery
   c) Inferior vena cava
   d) Lungs

9. Emphysema is:
   a) Inflammation of lung tissue
   b) Inflammation of the bronchial tubes
   c) Inflammation of the pleural lining of the lungs
   d) Loss of elasticity and stretch in alveoli

10. A collapsed lung is known as:
    a) Pneumonia
    b) Tuberculosis
    c) Pneumothorax
    d) Pleurisy

11. A genetic disease in which thick mucus clogs the lungs and other organs:
    a) Cor Pulmonale
    b) Sarcoidosis
    c) Chronic Obstructive Airways Disease
    d) Cystic Fibrosis

12. Inhaled air contains approximately the following percentage of oxygen:
    a) 59%
    b) 95%
    c) 21%
    d) 79%

13. Exhaled air contains approximately the following percentage of oxygen:
    a) 59%
    b) 95%
    c) 16%
    d) 79%

14. Inhaled air contains approximately the following percentage of nitrogen
    a) 59%
    b) 95%
    c) 21%
    d) 79%

Digestive System – circle correct answer

1. Where does ingestion begin?
   a) Stomach
   b) Mouth
   c) Small intestine
   d) Large intestine

2. Saliva amylase starts the breakdown of:
   a) Protein
   b) Fats
   c) Carbohydrates
   d) Minerals

3. Lipase digests
   a) Starch
   b) Carbohydrates
   c) Proteins
   d) Fats

4. Which of these nerves connects the brain and the gut:
   a) Trochlear nerve
   b) Vestibulocochlear nerve
   c) Phrenic nerve
   d) Vagus nerve

5. A bolus is a:
   a) Digestive enzyme
   b) Ball of food that leaves the mouth after chewing
   c) Ball of food that is trapped in the trachea
   d) Mineral found in food

6. The semifluid mixture of partly digested food is called:
   a) Chyme
   b) Chyle
   c) Colloids
   d) Cellulose

7. The acid secretion of the stomach is:
   a) Acetic acid
   b) Lactic acid
   c) Nucleic acid
   d) Hydrochloric acid

8. Food leaves the stomach by passing through the:
   a) Cardiac sphincter
   b) Pyloric sphincter
   c) Oesophageal sphincter
   d) Sphincter of Oddi

9. Bile is produced in the:
   a) Spleen
   b) Pancreas
   c) Liver
   d) Gallbladder

10. The villi in the small intestine contain lymphatic vessels, which absorb fats and they are called:
   a) Lacteals
   b) Lactones
   c) Lactates
   d) Lamellae

11. Chronic damage to an organ causing hardening of the organ is:
   a) An ulcer
   b) A hernia
   c) Cirrhosis
   d) Bulimia

12. Hepatitis is
   a) A fungal infection
   b) A viral infection
   c) An autoimmune condition
   d) A bacterial infection

13. Foods that serve as fertiliser for friendly bacteria in the gut:
   a) Probiotics
   b) Prebiotics
   c) Enzymes
   d) Minerals

14. This neurotransmitter communicates between the brain and the gut:
   a) Acetylcholine
   b) Dopamine
   c) Serotonin
   d) Noradrenaline

Nervous System – circle correct answer

1. A nerve impulse travels from:
   a) Dendrite to axon to cell body
   b) Axon to synapse to dendrite
   c) Cerebrum to cerebellum
   d) Cell body to synapse

2. The sympathetic nervous system affects the heart by:
   a) Accelerating its action
   b) Constricting blood flow to the heart
   c) Lowering blood pressure
   d) Slowing down its action

3. Multiple sclerosis is due to loss of:
   a) Neurotransmitters
   b) The myelin sheath
   c) Dendrites
   d) Grey matter

4. White matter is composed of:
   a) Arachnoid mater and pia mater
   b) Bundles of myelinated nerve fibres
   c) Cell bodies and unmyelinated axons & dendrites
   d) Pia mater and dura mater

5. The number of pairs of spinal and cranial nerves in the body is:
   a) 31 pairs of spinal nerves and 12 pairs of cranial nerves
   b) 28 pairs of spinal nerves and 10 pairs of cranial nerves
   c) 30 spinal nerves and 12 pairs of cranial nerves
   d) 31 spinal nerves and 14 pairs of cranial nerves

6. Cerebral Palsy affects:
   a) The motor system control
   b) The myelin sheath
   c) The intervertebral discs
   d) The parasympathetic nervous system

7. What is the function of a dendrite?
   a) To send nerve impulses to the cell body
   b) To send nerve impulses away from the cell body
   c) To pass the nerve impulse to the next neurone
   d) To be a chemical messenger

8. What is the function of a synapse?
   a) To insulate the axon
   b) To speed up the passage of the nerve impulse
   c) To pass on a nerve impulse to the axon of the next neurone
   d) To contain a chemical messenger, which enables a nerve impulse to be transmitted

9. The function of the olfactory cranial nerve is:
   a) To move the eyeballs
   b) Enable the sense of smell
   c) To supply the muscles of the tongue
   d) To enable the sense of sight

10. The outermost layer of the meninges is called:
    a) Pia mater
    b) Arachnoid mater
    c) Sub-arachnoid
    d) Dura mater

11. What effect does the sympathetic nervous system have on the skin?
    a) Dilation of blood vessels in the skin
    b) A decrease in ability of the skin to sweat
    c) Constriction of blood vessels in the skin
    d) Relaxation of erector pili muscles in the skin

12. Bell's Palsy is paralysis of which cranial nerve?
    a) Trigeminal
    b) Optic
    c) Facial
    d) Spinal accessory

13. The neurotransmitter for memory, learning and movement co-ordination:
    a) Acetylcholine
    b) Dopamine
    c) Serotonin
    d) Noradrenaline

14. Inhibits pain signals in the body:
    a) Dopamine
    b) Glutamate
    c) Adrenaline
    d) Endorphins

The Senses – circle correct answer

1. Thermoreceptors help detect:
    a) Pain
    b) Temperature
    c) Textures
    d) Smells

2. Receptors in the skin that detect pressure, vibrations, and texture:
    a) Thermoreceptor
    b) Proprioceptors
    c) Pain receptors
    d) Mechanoreceptors

3. Receptors that detect movement and position of the limbs:
    a) Thermoreceptor
    b) Proprioceptors
    c) Pain receptors
    d) Mechanoreceptors

4. Part of the brain involved with emotional association of smells & memory:
    a) Cerebrum
    b) Cerebellum
    c) Limbic system
    d) Pineal gland

5. Anosmia is:
    a) A heightened sense of smell
    b) A reduced sense of smell
    c) Ability to smell only certain odours
    d) Loss of sense of smell

6. The clear, thin layer covering the front of each eye is called:
    a) The iris
    b) The pupil
    c) The cornea
    d) The retina

7. This vitamin is important for the eyes:
    a) Vitamin E
    b) Vitamin A
    c) Vitamin B12
    d) Vitamin B6

8. Macular deterioration is caused by:
    a) Deterioration of part of the retina
    b) Deterioration of the iris
    c) Deterioration of the pupil
    d) Deterioration of the vitreous humor

9. The tympanic membrane is commonly called the:
    a) Ear canal
    b) Ear drum
    c) Middle ear
    d) Inner ear

10. The ossicles are:
    a) Tubes in the inner ear
    b) Three small bones in the middle ear
    c) Branches of the auditory nerve
    d) Sinus cavities

11. Age-related hearing loss is also known as:
    a) Tinnitus
    b) Glue ear
    c) Presbycusis
    d) Meniere's Disease

12. Ringing or buzzing in the ears in the absence of external sounds:
    a) Vertigo
    b) Meniere's disease
    c) Tinnitus
    d) Glue ear

13. Gustatory receptors detect:
    a) Touch
    b) Smells
    c) Tastes
    d) Sounds

14. Umami taste receptors detect:
    a) Sweet tastes
    b) Sour tastes
    c) Bitter tastes
    d) Savoury or meaty tastes

Endocrine System – circle correct answer

1. Oxytocin is a hormone secreted by the:
   a) Pituitary, anterior lobe
   b) Pituitary posterior lobe
   c) Adrenal cortex
   d) Adrenal medulla

2. Adrenocorticotrophic hormone (ACTH) controls the:
   a) Adrenal cortex
   b) Adrenal medulla
   c) Thyroid
   d) Height and growth

3. Grave's diseases is a result of:
   a) Thyroxin hyposecretion
   b) Thyroxin hypersecretion
   c) Thymic factor hyposecretion
   d) Thymic factor hypersecretion

4. Addison's Disease is usually caused by:
   a) Hyposecretion of adrenocortical hormones
   b) Hyposecretion of adrenal medulla hormones
   c) Hypersecretion of adrenocortical hormones
   d) Hypersecretion of adrenal medulla hormones

5. The hormone HGH regulates:
   a) Calcium levels in the blood
   b) Water absorption
   c) Height and growth
   d) Metabolism

6. These hormones are gonadotropins (they influence the sex glands):
   a) FSH, LH and ICSH
   b) FSH, LTH and ACTH
   c) LH, LTH and MSH
   d) ICSH, LH and LTH

7. Adrenal medulla hormones are elevated in cases of:
   a) Acromegaly
   b) Addison's disease
   c) PMS syndrome
   d) Stress

8. Hyposecretion of Vasopressin (ADH) can cause:
   a) Diabetes insipidus
   b) Oedema
   c) Hypertension
   d) Hypotension

9. The function of Interstitial Cell Stimulating Hormone (ICSH) is to:
   a) Stimulate sperm and testosterone production
   b) Regulate water absorption in kidneys
   c) Control body rhythms
   d) Metabolise carbohydrates

10. Levels of this hormone are often abnormal in people suffering from SAD (seasonal affective disorder):
    a) Melatonin
    b) Melanin
    c) ADH
    d) Thyroxin

11. Hashimoto's disease affects this endocrine organ:
    a) Pancreas
    b) Pituitary
    c) Thyroid
    d) Thymus

12. Hormone that increases calcium levels in blood when they are too low:
    a) Calcitonin
    b) Parathormone
    c) Thyroxin
    d) Adrenaline

13. Diabetes mellitus affects this endocrine organ:
    a) Pancreas
    b) Pituitary
    c) Thyroid
    d) Thymus

14. The islets of Langerhans are located here:
    a) Pituitary
    b) Thyroid
    c) Adrenals
    d) Pancreas

Urinary System – circle correct answer

1. Kidney tissue is made up of over a million twisted tubes called:
   a) Tubules
   b) Calyces
   c) Nephrons
   d) Capsules

2. This kidney hormone stimulates red blood cell production:
   a) Angiotensin
   b) Erythropoietin
   c) Renin
   d) Calcitriol

3. Initial collection point in kidney nephron for waste products in the blood:
   a) The renal vein
   b) The loop of Henle
   c) Bowman's capsule
   d) Renal Pelvis

4. Reabsorption of substances that are not waste takes place in the:
   a) The tubules
   b) Glomerulus
   c) Renal pelvis
   d) Ureters

5. Approximate amount of urine, which causes the bladder to contract:
   a) 1,000ml
   b) 500ml
   c) 200ml
   d) 100ml

6. Approximate amount of urine produced by average adult every 24 hours:
   a) 1.5 litres
   b) 10 litres
   c) 5 litres
   d) 0.5 litre

7. The function of the ureters is to:
   a) Collect and store urine
   b) Take urine from the bladder to the outside of the body
   c) Filter urine
   d) Take urine from the kidneys to the bladder

8. Women are more susceptible to bladder infections than men because:
   a) They have larger bladders
   b) The female urethra is shorter
   c) Female urine is more acidic
   d) Female ureters are shorter

9. The need to urinate frequently during the night is called:
   a) Oliguria
   b) Anuria
   c) Haematuria
   d) Nocturia

10. Pregnant women are screened for this, as it may signal a condition called preeclampsia:
    a) Glucose in the urine
    b) Salts in the urine
    c) Protein in the urine
    d) Blood in the urine

11. Which organ can sometimes cause urinary problems in men?
    a) Spleen
    b) Prostate
    c) Cowper's glands
    d) Bulbourethral glands

12. Kidney problems are commonly caused by:
    a) High blood pressure
    b) Low blood pressure
    c) Low cholesterol
    d) High cholesterol

13. Treatment that filters and purifies the blood using a machine:
    a) Diastole
    b) Dialysis
    c) Diastasis
    d) Machine filtration

14. Production of unusually small amounts of urine:
    a) Oliguria
    b) Anuria
    c) Haematuria
    d) Nocturia

Reproductive System – circle correct answer

1. The neck of the uterus is called the:
   a) Renal pelvis
   b) Cervix
   c) Sphincter
   d) Caecum

2. The ovaries produce:
   a) Luteinising hormone
   b) Gonadotropin
   c) Oestrogen
   d) Testosterone

3. The innermost lining of the uterus is called the:
   a) Endomorph
   b) Embolism
   c) Epimetrium
   d) Endometrium

4. In a 28-day female menstrual cycle, ovulation is likely to occur around day:
   a) 1
   b) 7
   c) 14
   d) 28

5. Fertilisation of the ovum takes place in:
   a) The ovary
   b) The fallopian tube
   c) The uterus
   d) The vagina

6. The period of time from fertilisation to birth is called:
   a) Gastrulation
   b) Menstruation
   c) Germination
   d) Gestation

7. Where would you find a Graafian follicle?
   a) On the surface of a fallopian tube
   b) On the surface of an ovary
   c) Inside the uterus
   d) Inside the fallopian tubes

8. Dysmenorrhea can be defined as:
   a) Extremely painful menstruation
   b) Absence of menstruation
   c) Irregular menstruation
   d) Heavy menstruation

9. Hyperemesis gravidarum is:
   a) Heavy menstrual periods
   b) Extreme morning sickness during pregnancy
   c) Placenta partially covering the cervix
   d) Gestational diabetes

10. Fusion of sperm and ovum creates a:
    a) Blastocyst
    b) Morula
    c) Zygote
    d) Foetus

11. Sperm are stored and mature in the:
    a) Vas deferens
    b) Epididymis
    c) Seminal vesicles
    d) Cowper's glands

12. The pouch-like sac of skin that holds the testes is called the:
    a) Scrotum
    b) Epididymis
    c) Seminal vesicles
    d) Cowper's glands

13. Benign prostatic hyperplasia (BPH) is:
    a) Inflammation of the prostate gland
    b) Reduction in size of the prostate gland
    c) Enlargement of the prostate gland
    d) Cancer of the prostate gland

14. How many chromosomes do an ovum and sperm each contain?
    a) 46
    b) 23
    c) 35
    d) 40

Mixed Multiple Choice Test 1 – circle correct answer

1. This non-contagious skin condition causes erythema on the face:
   a) Herpes simplex
   b) Herpes zoster
   c) Rosacea
   d) Verruca

2. This skin condition causes itching:
   a) Milia
   b) Urticaria
   c) Comedones
   d) Xanthomas

3. This is an inflammation of body tissues, especially below the skin:
   a) Cellulitis
   b) Cellulite
   c) Acne vulgaris
   d) Naevae

4. This smooth bluish-white tissue is found on the articular surface of joints:
   a) Yellow elastic cartilage
   b) Fibrocartilage
   c) Hyaline cartilage
   d) Adipose tissue

5. This tissue is involved in immunity:
   a) Areolar
   b) Adipose
   c) White fibrous
   d) Lymphoid

6. This condition is also called shingles:
   a) Vitiligo
   b) Chloasma
   c) Herpes zoster
   d) Herpes simplex

7. Sebaceous glands secrete:
   a) Sebum
   b) Perspiration
   c) Lymph
   d) Blood

8. Foul-smelling perspiration is medically termed as:
   a) Hyperhidrosis
   b) Anhidrosis
   c) Bromidrosis/osmidrosis
   d) Carcinoma

9. 'Hallux' refers to which part of the body:
   a) Arm
   b) Hand
   c) Little toe
   d) Big toe

10. The action of the Pectoralis Major muscle is to:
    a) Rotate arm outwards
    b) Elevate shoulder and rotate scapula
    c) Abduct arm
    d) Draw arm forward, adduct and rotate arm inwards

11. This muscle is located on the upper arm:
    a) Tibialis anterior
    b) Flexor digitorum
    c) Brachialis
    d) Flexor carpi ulnaris

12. This muscle flexes the neck:
    a) Levator scapula
    b) Sternocleidomastoid
    c) Rhomboids
    d) Splenius cervicus

13. This type of tissue is composed of fat cells:
    a) Adipose
    b) Areolar
    c) Stratified keratinized
    d) Connective

14. Osteoporosis is:
    a) A progressive disease causing brittle bones prone to fractures
    b) A genetic disease in which the bones are abnormally brittle
    c) A chronic disease causing bones to become abnormally enlarged and weak
    d) Softening of the bones due to a deficiency of Vitamin D

15. Where would you find a hinge joint?
    a) Between radius & ulna
    b) Between tarsals
    c) Sacro-iliac joint
    d) Between phalanges

16. This is a foot bone:
    a) Calcaneus
    b) Ulna
    c) Carpal
    d) Radius

17. The mandible is another name for the:
    a) Upper jaw
    b) Cheek
    c) Nose
    d) Lower jaw

18. An effect of ankylosing spondylitis is:
    a) Cartilage in joints breaks down
    b) Chronic destruction of synovial membrane
    c) Fusion of joints in the spine
    d) Bones become brittle and porous

19. How many pairs of ribs do humans have?
    a) Three
    b) Seven
    c) Twelve
    d) Two

20. This is a common effect of diabetes mellitus:
    a) Depression, bloating
    b) Fight or flight
    c) Frequent urination
    d) Painful periods

21. The nervous & endocrine systems work together to ensure:
    a) Metabolism
    b) Body imbalance
    c) Control of menstruation
    d) Homeostasis

22. What is the function of parathormone?
    a) To metabolise carbohydrates, proteins & fats
    b) To metabolise vitamins & minerals
    c) To regulate calcium levels in the blood
    d) To regulate water absorption

23. These blood vessels carry oxygenated blood away from the lungs & back to the heart:
    a) Pulmonary veins
    b) Pulmonary arteries
    c) Aorta
    d) Superior vena cava

24. How does oxygen enter the blood stream?
    a) Osmosis
    b) Active transport
    c) Diffusion
    d) Dissolution

25. This part of the brain is involved with intelligence and emotions:
    a) Cerebellum
    b) Hypothalamus
    c) Cerebrum
    d) Medulla

26. Part of a nerve cell that transmits signals away from the cell body:
    a) Dendrite
    b) Axon
    c) Synapse
    d) Myelin

27. The epididymis is located:
    a) At the base of each testicle
    b) On top of each testicle
    c) In the fallopian tubes
    d) In the prostate gland

28. A benign tumour in the uterus is called:
    a) Endometriosis
    b) A zygote
    c) A blastocyst
    d) A fibroid

29. This is the first site of carbohydrate digestion where polysaccharides are broken down:
    a) Mouth
    b) Oesophagus
    c) Large intestine
    d) Small intestine

30. The muscle that allows food to exit the stomach is called the:
    a) Cardiac sphincter
    b) Pyloric sphincter
    c) Anal sphincter
    d) Abdominal sphincter

31. This part of the nervous system triggers expiration (breathing out) in the respiratory system:
    a) Pons
    b) Medulla
    c) Cerebrum
    d) Cerebellum

32. The nerve that controls the diaphragm is called:
    a) Sciatic Nerve
    b) Trigeminal nerve
    c) Phrenic nerve
    d) Vagus nerve

33. This is a fungal infection:
    a) Rosacea
    b) Alopecia areata
    c) Pediculosis capitis
    d) Tinea pedis

34. The most common form of skin cancer is called:
    a) Melanoma
    b) Basal cell carcinoma
    c) Squamous cell carcinoma
    d) Chloasma

35. The most deadly form of skin cancer is called:
    a) Melanoma
    b) Basal cell carcinoma
    c) Squamous cell carcinoma
    d) Chloasma

36. Multiple sclerosis is due to damage to:
    a) Neurotransmitters
    b) Glial cells
    c) Myelin sheath of nerve axons
    d) Nodes of Ranvier

37. Where would you find the subclavian veins?
    a) Neck
    b) Head
    c) Abdomen
    d) Leg

38. The prostate gland is located:
    a) At the upper part of the urethra
    b) Above the bladder
    c) Behind the rectum
    d) On top of each testis

39. What organ do ulcers commonly affect?
    a) Duodenum
    b) Ovaries
    c) Uterus
    d) Bladder

40. The endometrium is another name for the:
    a) Lining of the lungs
    b) Lining of the heart
    c) Lining of the bladder
    d) Lining of the uterus

41. What organ does nephritis affect?
    a) Bladder
    b) Kidney
    c) Prostate
    d) Uterus

42. This hormone is derived from serotonin:
    a) ICSH
    b) FSH
    c) Melatonin
    d) Glucagon

43. Enlargement of bones in hands, feet & face, often accompanied by headache are symptoms of:
    a) Addison's disease
    b) Cushing's syndrome
    c) Gigantism
    d) Acromegaly

44. FSH & LH affect which organs?
    a) Pancreas
    b) Ovaries
    c) Uterus
    d) Pituitary

45. Childbirth usually occurs about how many weeks after fertilization?
    a) 30
    b) 35
    c) 40
    d) 45

46. A pregnancy that begins to develop outside the uterus is called:
    a) Ectopic
    b) Embryonic
    c) Prolapse
    d) Endometriosis

47. The Graafian follicle is the name given to a:
    a) Undeveloped follicle
    b) Dormant follicle
    c) Maturing follicle
    d) Ruptured follicle

48. An obstruction of lymphatic flow results in:
    a) Lymphoma
    b) Cellulite
    c) Lymphadenitis
    d) Oedema

49. The largest lymphatic vessel is called:
    a) Thoracic duct
    b) Right lymphatic duct
    c) Inferior vena cava
    d) Superior vena cava

50. Inflammation of a nerve is called:
   a) Neuralgia
   b) Neuritis
   c) Neuropathy
   d) Neurotransmitter

Word Search 1 (Horizontal & Vertical)

D G O I T R E D D P B B Y A A G P

J P W E L O G O U N W R D C N R Q

C G I N G I V I T I S O I R E A A

H A D L B T F G O U T N A O U L Z

O R E P U G N B X V V C B M R O E

L T M O R H U U I E K H E E Y P M

E H E D S A Y L E C U I T G S E P

C R N O I J F I O Z Y T E A M C H

Y I T K T E E M H E Y I S L Z I Y

S T I J I Z V I W M R S I Y T A S

T I A K S E E A A A C A N C E R E

I S W C I R R H O S I S X K I B M

T E N D O M E T R I O S I S H U A

I K S C Y S T I T I S U S N J N S

S I F K C Z G D E R M A T I T I S

E P I L E P S Y P T Q X M B R O D

D I V E R T I C U L I T I S N N O

gingivitis gout goitre fever endometriosis
emphysema epilepsy eczema dermatitis diverticulitis
diabetes dementia cystitis cirrhosis cholecystitis
cancer bulimia bursitis bunion bronchitis arthritis
aneurysm alopecia acromegaly

Mixed Multiple Choice Test 2 – circle correct answer

1. Pathology refers to:
   a) The mechanism of disease
   b) The breakdown of compounds
   c) Using oxygen to produce energy
   d) Joining two different compounds together

2. Which of the following stores DNA inside the cell?
   a) Nucleus
   b) Golgi apparatus
   c) Ribosomes
   d) Mitochondria

3. Which of the opposite anatomical terms is INCORRECT:
   a) Superior - Inferior
   b) Superficial - Deep
   c) Proximal - Medial
   d) Dorsal - Ventral

4. Which of the following is one of the major cavities of the body:
   a) Brachial
   b) Femoral
   c) Thoracic
   d) Axial

5. How many chromosomes are inside a reproductive cell (sperm, ova)?
   a) 46
   b) 36
   c) 23
   d) 56

6. Which of the following is involved in producing energy inside the cell?
   a) Nucleus
   b) Golgi apparatus
   c) Ribosomes
   d) Mitochondria

7. Where are long bones located?
   a) Thorax
   b) Limbs
   c) Skull
   d) Pelvis

8. Some flat bones are located here:
   a) Wrist
   b) Limbs
   c) Skull
   d) Spine

9. Which bone is part of the axial skeleton?
   a) Humerus
   b) Ulna
   c) Rib
   d) Tibia

10. Which bone is part of the appendicular skeleton?
   a) Tibia
   b) Maxilla
   c) Sternum
   d) Hyoid

11. What is kyphosis of the spine?
   a) Increased lumbar curve
   b) Reduced lumbar curve
   c) Increased thoracic curve
   d) Reduced thoracic curve

12. Paget's Disease involves:
   a) Increased bone strength
   b) Decreased blood flow to bone
   c) Increased bone cell activity
   d) Decreased bone cell activity

13. Ligaments:
   a) Join muscle to bone
   b) Join tendon to bone
   c) Join bone to bone
   d) Join tendon to muscle

14. Which of the following describes movement TOWARDS the midline?
   a) Adduction
   b) Flexion
   c) Inversion
   d) Supination

15. The hip and shoulder joints are examples of:
    a) Gliding joints
    b) Ball and socket joints
    c) Hinge Joints
    d) Saddle joints

16. The epidermis is made up of this type of tissue:
    a) Cuboidal epithelium
    b) Ciliated epithelium
    c) Stratified squamous epithelium
    d) Transitional epithelium

17. Which term describes a deep open lesion on the skin, which penetrates all 3 layers?
    a) Fissure
    b) Excoriation
    c) Ulcer
    d) Abrasion

18. Stretchmarks involve:
    a) Tears in the dermis
    b) Tears in the epidermis
    c) Tears in blood vessels
    d) Tears in hair follicles

19. Immediate allergic reactions involve a chemical called:
    a) Gamma Butyric Acid
    b) Dopamine
    c) Glutamate
    d) Histamine

20. Suspicious features of skin moles include:
    a) Irregular borders
    b) 2 or more colours
    c) Bleeding
    d) All of the above

21. Skeletal muscle is:
    a) Under voluntary control
    b) Under involuntary control
    c) Also called visceral muscle
    d) Found in the intestinal wall

22. Cardiac muscle is:
    a) Under voluntary control
    b) Smooth
    c) Found in blood vessels
    d) NONE of the above

23. The tibiofemoral joint is a:
    a) Hinge joint
    b) Saddle joint
    c) Ball and socket joint
    d) Pivot joint

24. A muscle's ORIGIN is:
    a) Attachment of muscle tendon to the moveable bone
    b) Attachment of muscle tendon between two bones
    c) Attachment of muscle tendon to the stationary bone
    d) None of the above

25. The rotator cuff muscles are comprised of the:
    a) Teres minor, Supraspinatus, Infraspinatus, Subscapularis
    b) Teres major, Teres minor, Infraspinatus, Subscapularis
    c) Teres minor, Supraspinatus, Infraspinatus, Serratus posterior
    d) Teres major, Teres minor, Supraspinatus, Infraspinatus

26. Which of the following filter blood:
    a) Spleen
    b) Liver
    c) Kidney
    d) All of the above

27. The deepest layer of muscle in the abdominal wall is:
    a) Transversus abdominis
    b) Internal obliques
    c) Rectus abdominis
    d) External obliques

28. How many groups of fibres does the deltoid muscle have?
    a) 1
    b) 2
    c) 3
    d) 4

29. The nerve that can be compressed by a tight piriformis muscle is:
    a) Median
    b) Sciatic
    c) Cubital
    d) Ulnar

30. The quadriceps muscles:
    a) Flex the knee
    b) Laterally rotate the hip
    c) Extend the hip
    d) Extend the knee

31. The branch of the nervous system that is involved in digestion:
    a) Sympathetic
    b) Somatic
    c) Central
    d) Parasympathetic

32. This region of the brain is involved in maintaining posture and balance:
    a) Cerebrum
    b) Cerebellum
    c) Pons
    d) Medulla

33. The 'thinking & problem solving' area of the cerebrum is in the:
    a) Frontal lobe
    b) Parietal lobe
    c) Occipital lobe
    d) Temporal lobe

34. Lacteals in the lymphatic system absorb:
    a) Carbohydrates
    b) Fats
    c) Protein
    d) Hormones

35. The natural 'pacemaker' of the heart (the sino-atrial node) is located in:
    a) The left atrium
    b) The right atrium
    c) The right ventricle
    d) The left ventricle

36. The LEFT atrium of the heart:
    a) Receives deoxygenated blood from the lungs
    b) Receives oxygenated blood from the ventricles
    c) Receives deoxygenated blood from the body
    d) Receives oxygenated blood from the lungs

37. Which of the following are associated with risk of heart attack and stroke?
    a) Diabetes
    b) High Cholesterol
    c) Smoking
    d) ALL of the above

38. The major artery leaving the left side of the heart is:
    a) The vena cava
    b) The aorta
    c) The pulmonary artery
    d) The coronary artery

39. This disease damages the villi in the small intestine:
    a) Crohn's disease
    b) Addison's disease
    c) Grave's disease
    d) Coeliac disease

40. The cartilage flap that closes the airway during swallowing is:
    a) The trachea
    b) The tongue
    c) The epiglottis
    d) The vocal cord

41. Which of the following statements is FALSE:
    a) The trachea is commonly called the windpipe
    b) The trachea lies in front of the oesophagus
    c) The trachea leads directly into the lungs
    d) The trachea contains cartilage

42. Which of the following statements is TRUE:
    a) Bronchial veins provide nutrition for the lungs
    b) Gas exchange occurs in the lungs between the alveoli and capillaries
    c) Gas exchange occurs between bronchi and bronchioles
    d) None of the above is true

43. The muscle that is critically important for breathing is:
    a) Rectus abdominis
    b) External obliques
    c) Diaphragm
    d) Internal obliques

44. Bile:
    a) Aids protein breakdown
    b) Aids carbohydrate breakdown
    c) Aids nucleic acid breakdown
    d) Aids fat breakdown

45. Which hormone pair is involved in glucose control?
    a) Thyroxin-Calcitonin
    b) Cortisol-Oxytocin
    c) Melatonin-Histamine
    d) Glucagon-Insulin

46. Which of these hormones is produced by the anterior pituitary gland?
    a) Cortisol
    b) Vasopressin
    c) Oxytocin
    d) Thyroid stimulating hormone

47. Vasopressin:
    a) Conserves body water (antidiuretic effect)
    b) Regulates blood glucose
    c) Stimulates uterine contraction during labour
    d) Raises metabolic rate

48. Salivary amylase begins the breakdown of?
    a) Carbohydrates
    b) Protein
    c) Fats
    d) None of the above

49. Which hormone is released during times of prolonged stress?
    a) Cortisol
    b) Melatonin
    c) Oxytocin
    d) Prolactin

50. Endometriosis:
   a) Usually presents in puberty
   b) Commonly involves the lungs
   c) Involves growth of the uterine lining outside the uterus
   d) Is caused by obesity

Word Search 2 (Horizontal & Vertical)

Y I G L A H L U M B A G O K L J H
A C E I D I N E M M O D K L E R A
I K M M K V E A A L I X E N U M E
J C I E V E P H S G P T N I K E M
L I G N U S H Y T M L H E N A N O
Y M R I H P R P I E A Y U S E O R
M P A N E K I O T N R P R O M R R
P E I G R E T T I O Y E I M I R H
H T N I N Q I H S P N R T N A H O
O I E T I G S Y C A G T I I N A I
M G C I A J N R Y U I E S A X G D
A O J S Y V F O G S T N A G M I S
V I J A U N D I C E I S C J U A G
U A L H P B U D F P S I N B M L A
N E U R A L G I A A N O E V P W W
G P L V N A U S E A B N N L S Z P
K O V H M Q G M X I M E A S L E S

acne haemorrhoids hernia hives hypertension
hypothyroidism impetigo insomnia jaundice laryngitis
leukaemia lumbago lymphoma mastitis measles
meningitis menopause menorrhagia migraine mumps
nausea nephritis neuralgia neuritis

Mixed Multiple Choice Test 3 – circle correct answer

1. Mitosis is the process of:
   a) Eliminating toxins from the skin
   b) Cell division
   c) Using oxygen to produce energy
   d) Joining two different compounds together

2. The functions of areolar tissue include:
   a) Protection, insulation
   b) Lining, absorption
   c) Support, connection
   d) Stretch, recoil

3. The liver is found in which quadrant of the body?
   a) Right lower
   b) Right upper
   c) Left lower
   d) Left upper

4. The term 'distal' means:
   a) Closer to the trunk
   b) Further from the trunk
   c) On the left side
   d) On the right side

5. How many chromosomes are inside a sex cell (ovum or sperm)?
   a) 26
   b) 23
   c) 46
   d) 56

6. What type of infection is acne vulgaris?
   a) Viral
   b) Congenital
   c) Bacterial
   d) Fungal

7. The cells that produce collagen & elastin are called:
   a) Goblet cells
   b) Macrophages
   c) Mast cells
   d) Fibroblasts

8. Where are short bones located?
   a) Wrist
   b) Limbs
   c) Skull
   d) Spine

9. Which bone is part of the appendicular skeleton?
   a) Humerus
   b) Sacrum
   c) Coccyx
   d) Ribs

10. Which bone is part of the axial skeleton?
    a) Tibia
    b) Calcaneus
    c) Sternum
    d) Fibula

11. What is lordosis of the spine?
    a) Increased lumbar curve
    b) Reduced lumbar curve
    c) Increased thoracic curve
    d) Reduced thoracic curve

12. Gout involves:
    a) Excess uric acid build up
    b) Inadequate uric acid
    c) Increased bone cell activity
    d) Decreased bone cell activity

13. The end of a muscle which moves least during contraction is called the:
    a) Insertion
    b) Tendon
    c) Origin
    d) Ligament

14. Which of the following describes movement AWAY FROM the midline?
    a) Adduction
    b) Abduction
    c) Inversion
    d) Supination

15. Joint between the atlas and axis of the cervical spine is an example of a:
    a) Pivot joint
    b) Ball and socket joint
    c) Hinge joint
    d) Saddle joint

16. The skin, hair & nails are collectively called:
    a) The Integumentary system
    b) Ciliated epithelium
    c) Stratified squamous epithelium
    d) Transitional epithelium

17. Carpal tunnel syndrome affects which joint?
    a) Ankle
    b) Sacroiliac
    c) Wrist
    d) Elbow

18. RICE (recommended for sports injuries) stands for:
    a) Recovery, Injections, Compression, Elevation
    b) Rest, Ice, Compression, Elevation
    c) Running, Ice, Care, Exercise
    d) Repair, Ice, Compression, Elevation

19. An action of the biceps femoris muscle is to:
    a) Extend the knee
    b) Abduct the leg
    c) Flex the knee
    d) Flex the elbow

20. If muscles remain inactive for long periods the following can occur:
    a) Cramp
    b) Atrophy
    c) Lactic acid build up
    d) Hypertrophy

21. Smooth muscle is:
    a) Also called visceral muscle
    b) Under involuntary control
    c) Found in the intestinal wall
    d) All of the above

22. Plantar fasciitis affects:
    a) Top of the foot
    b) Sole of the foot
    c) Palm of the hand
    d) Calf muscles

23. The patellofemoral joint is a:
    a) Hinge joint
    b) Saddle joint
    c) Gliding joint
    d) Pivot joint

24. When muscle cells don't get enough oxygen this can result:
    a) Uric acid build up
    b) Lactic acid build up
    c) Citric acid build up
    d) Hydrochloric acid build up

25. The quadriceps muscles are comprised of:
    a) Rectus Femoris, Supraspinatus, Infraspinatus, Vastus Lateralis
    b) Teres Major, Teres Minor, Vastus Intermedius, Rectus Femoris
    c) Rectus Femoris, Vastus Medialis, Vastus Lateralis, Vastus Intermedius
    d) Teres Major, Teres Minor, Supraspinatus, Infraspinatus

26. Which of the following organs do not filter blood:
    a) Spleen
    b) Liver
    c) Kidney
    d) Pancreas

27. The most superficial layer of muscle in the abdominal wall is:
    a) Transversus abdominis
    b) Internal obliques
    c) Rectus abdominis
    d) External obliques

28. How many hamstrings muscles are there?
    a) 1
    b) 2
    c) 3
    d) 4

29. The nerve that controls the chewing muscles and carries pain signals from the face is called:
    a) Vagus
    b) Facial
    c) Trigeminal
    d) Spinal accessory

30. The Gastrocnemius & Soleus muscles:
    a) Flex the knee
    b) Plantarflex the foot
    c) Flex the toes
    d) Extend the knee

31. The branch of the nervous system, which slows down the action of the heart, is called:
    a) Sympathetic
    b) Somatic
    c) Central
    d) Parasympathetic

32. White matter is mainly composed of:
    a) Nerve cell bodies
    b) Myelinated nerve fibres (axons)
    c) Myelinated nerve fibres (dendrites)
    d) Neuroglia

33. The hearing area of the brain is located in the:
    a) Frontal lobe
    b) Parietal lobe
    c) Occipital lobe
    d) Temporal lobe

34. Used by the body for growth and repair:
    a) Carbohydrate
    b) Fat
    c) Protein
    d) Water

35. Oxygenated blood from the lungs first enters:
    a) The left atrium
    b) The right atrium
    c) The right ventricle
    d) The left ventricle

36. The RIGHT atrium of the heart:
    a) Receives deoxygenated blood from the lungs
    b) Receives oxygenated blood from the ventricles
    c) Receives deoxygenated blood from the body
    d) Receives oxygenated blood from the lungs

37. An inherited disorder in which the blood clots very slowly is called:
    a) Septicaemia
    b) Haemophilia
    c) Thrombophilia
    d) Thrombosis

38. An aneurysm is:
    a) A build up of fats inside a blood vessel
    b) A gradual hardening of a blood vessel
    c) A weakening of an area of a blood vessel causing a localised dilation
    d) Blockage of a coronary artery

39. Cerebrospinal fluid is found:
    a) Around the brain only
    b) Between the meninges and the vertebrae of the spine
    c) Within the meninges of the brain & spinal cord
    d) Outside the meninges of the brain & spinal cord

40. The vocal cords are located in:
    a) The trachea
    b) The larynx
    c) The epiglottis
    d) The pharynx

41. Emphysema is irreversible damage to which part of the respiratory system?
    a) Trachea
    b) Pharynx
    c) Alveoli
    d) Bronchioles

42. Which of the following statements is TRUE:
    a) Cystic fibrosis is a genetic disease
    b) Cystic fibrosis is a disease causing lack of mucous in the body
    c) Cystic fibrosis affects the muscles and bones
    d) Cystic fibrosis is also called seasonal allergic rhinitis

43. The muscle that extends the elbow is called:
    a) Biceps
    b) Deltoid
    c) Serratus anterior
    d) Triceps

44. Amylase:
    a) Aids protein breakdown
    b) Aids carbohydrate breakdown
    c) Aids nucleic acid breakdown
    d) Aids fat breakdown

45. Which hormone pair is involved in calcium control?
    a) Parathormone-Calcitonin
    b) Cortisol-Oxytocin
    c) Melatonin-Histamine
    d) Glucagon-Insulin

46. Which of the following hormones is produced by the adrenal medulla?
    a) Cortisol
    b) Vasopressin
    c) Adrenaline
    d) Thymosin

47. Oxytocin:
    a) Conserves body water (antidiuretic effect)
    b) Regulates blood glucose
    c) Stimulates uterine contraction during labour
    d) Raises metabolic rate

48. Over production of human growth hormone during adulthood results in a condition called:
    a) Gigantism
    b) Cretinism
    c) Cushing's syndrome
    d) Acromegaly

49. Inflammation of the bladder is called:
    a) Cystitis
    b) Oliguria
    c) Haematuria
    d) Renal colic

50. Stein-Levental syndrome is commonly called:
    a) Endometriosis
    b) Fibroids
    c) Painful periods
    d) Polycystic ovaries

Word Search 3 (Horizontal & Vertical)

I	P	X	O	Y	P	L	E	U	R	I	S	Y	I	K	Z	R
O	T	I	T	I	S	S	W	L	C	E	K	N	M	F	W	O
O	S	H	I	N	G	L	E	S	C	A	B	I	E	S	O	S
S	N	E	U	R	A	L	G	I	A	P	C	K	H	I	S	A
T	S	K	R	I	C	K	E	T	S	O	L	X	B	M	T	C
E	J	J	J	W	I	S	C	O	L	I	O	S	I	S	E	E
O	H	L	O	S	T	E	O	P	O	R	O	S	I	S	O	A
M	L	O	R	D	O	S	I	S	E	P	S	I	S	C	A	K
A	I	P	S	O	R	I	A	S	I	S	O	E	H	R	R	Y
L	P	S	C	I	A	T	I	C	A	S	E	L	U	H	T	P
A	H	I	Y	V	L	N	B	F	F	L	D	G	B	I	H	H
C	P	S	I	N	U	S	I	T	I	S	E	E	Q	N	R	O
I	C	K	R	I	N	G	W	O	R	M	M	F	D	I	I	S
A	S	E	P	T	I	C	A	E	M	I	A	K	N	T	T	I
L	Z	H	P	R	O	S	T	A	T	I	T	I	S	I	I	S
P	D	P	N	E	U	M	O	N	I	A	D	P	D	S	S	B
J	G	N	P	A	L	P	I	T	A	T	I	O	N	S	B	C

kyphosis lordosis neuralgia oedema osteoarthritis

osteomalacia osteoporosis otitis palpitations pleurisy

pneumonia prostatitis psoriasis rhinitis rickets

ringworm rosacea scabies sciatica scoliosis sepsis

septicaemia shingles sinusitis

Mixed Multiple Choice Test 4 – circle correct answer

1. Catabolism refers to:
   a) The mechanism of disease
   b) The breakdown of compounds
   c) Using oxygen to produce energy
   d) Joining two different compounds together

2. Which of the following produces energy inside the cell?
   a) Nucleus
   b) Golgi apparatus
   c) Ribosomes
   d) Mitochondria

3. Which pair of opposite anatomical terms is INCORRECT?
   a) Superior - Inferior
   b) Superficial - Deep
   c) Proximal - Distal
   d) Dorsal - Posterior

4. Which of the following is one of the cavities of the body:
   a) Brachial
   b) Femoral
   c) Cranial
   d) Axial

5. An extra one of these chromosomes is found in Down syndrome:
   a) 23
   b) 12
   c) 21
   d) 14

6. Which of the following is involved in producing protein inside the cell?
   a) Nucleus
   b) Golgi apparatus
   c) Ribosomes
   d) Mitochondria

7. These are categorised as long bones:
   a) Carpals
   b) Phalanges
   c) Skull bones
   d) Vertebrae

8. Bones classified as irregular shaped bones are found here:
   a) Wrist
   b) Limbs
   c) Fingers
   d) Spine

9. Which bone is part of the axial skeleton?
   a) Humerus
   b) Ulna
   c) Sternum
   d) Tibia

10. Which bone is part of the appendicular skeleton?
    a) Fibula
    b) Maxilla
    c) Sternum
    d) Hyoid

11. What is scoliosis of the spine?
    a) Increased lumbar curve
    b) Reduced lumbar curve
    c) Increased thoracic curve
    d) Sideways curvature of the spine

12. Osteoporosis involves:
    a) Increased bone strength
    b) Decreased blood flow to bone
    c) Increased bone cell activity
    d) Bones becoming brittle and fragile

13. Periosteum:
    a) Joins muscle to bone
    b) Joins tendon to bone
    c) Join bone to bone
    d) Covers surface of bone

14. Which of the following describes movement TOWARDS the midline?
    a) Abduction
    b) Flexion
    c) Adduction
    d) Supination

15. The carpals and tarsals have:
    a) Gliding joints
    b) Ball and socket joints
    c) Hinge Joints
    d) Saddle joints

16. This is found in the upper respiratory system:
    a) Cuboidal epithelium
    b) Ciliated epithelium
    c) Stratified squamous epithelium
    d) Transitional epithelium

17. Which term describes a burn on the skin, which penetrates all 3 layers?
    a) Fourth degree burn
    b) Third degree burn
    c) Second degree burn
    d) First degree burn

18. Skin disease caused by hardening and thickening of the dermis:
    a) Striae
    b) Discoid lupus
    c) Scleroderma
    d) Folliculitis

19. This neurotransmitter is associated with wellbeing:
    a) Serotonin
    b) Dopamine
    c) Glutamate
    d) Noradrenaline

20. Skin moles that change and grow may indicate:
    a) Basal cell carcinoma
    b) Melanoma
    c) Squamous cell carcinoma
    d) All of the above

21. Skeletal muscle is:
    a) Controlled by the somatic nervous system
    b) Under involuntary control
    c) Also called visceral muscle
    d) Found in the intestinal wall

22. Smooth muscle is
    a) Under voluntary control
    b) Attached to skeletal bones
    c) Found in blood vessels and internal organs
    d) Also called cardiac muscle

23. The hip joint is a:
    a) Hinge joint
    b) Saddle joint
    c) Ball and socket joint
    d) Pivot joint

24. This type of muscle contraction does not change the muscle's length:
    a) Isotonic
    b) Isometric
    c) Concentric
    d) Eccentric

25. The rotator cuff muscles are located in the:
    a) Shoulder
    b) Neck
    c) Hip
    d) Foot

26. Which of the following organs stores and releases bile:
    a) Spleen
    b) Gall bladder
    c) Kidney
    d) Pancreas

27. Adbominal muscle referred to as a '6 pack' when well developed:
    a) Transversus abdominis
    b) Internal obliques
    c) Rectus abdominis
    d) External obliques

28. The intercostal muscles are found?
    a) In the pelvis
    b) Between the ribs
    c) In the abdomen
    d) In the lower limbs

29. This nerve can be compressed in carpal tunnel syndrome:
    a) Median
    b) Sciatic
    c) Cubital
    d) Ulnar

30. The hamstring muscles:
    a) Flex the knee
    b) Laterally rotate the hip
    c) Extend the elbow
    d) Extend the knee

31. The branch of the nervous system that is involved in relaxation:
    a) Sympathetic
    b) Somatic
    c) Central
    d) Parasympathetic

32. This region of the brain is involved with respiration:
    a) Cerebrum
    b) Cerebellum
    c) Limbic system
    d) Pons

33. The auditory cortex of the cerebrum is in the:
    a) Frontal lobe
    b) Parietal lobe
    c) Occipital lobe
    d) Temporal lobe

34. A sluggish lymphatic system can cause:
    a) Bleeding
    b) Oedema
    c) Weight loss
    d) Hearing loss

35. Disease of the heart, affecting its size, shape and structure is called:
    a) Cardiac failure
    b) Cardiomyopathy
    c) Bradycardia
    d) Tachycardia

36. Regarded as normal blood pressure in a healthy, resting adult:
    a) 120/80
    b) 150/100
    c) 160/80
    d) 80/120

37. Which of the following is commonly associated with risk of peripheral neuropathy?
    a) Diabetes
    b) High Cholesterol
    c) Weight gain
    d) Pregnancy

38. The major artery leaving the left side of the heart is:
    a) The vena cava
    b) The aorta
    c) The pulmonary artery
    d) The coronary artery

39. This disease causes inflammation in the digestive system:
    a) Crohn's disease
    b) Addison's disease
    c) Grave's disease
    d) Coeliac disease

40. The larynx includes:
    a) The trachea
    b) The tonsils
    c) The alveoli
    d) The vocal cords

41. Which of the following statements is FALSE:
    a) The trachea is also called the windpipe
    b) The trachea lies in front of the oesophagus
    c) The trachea leads directly into the lungs
    d) The trachea contains cartilage

42. Which of the following statement is TRUE:
    a) Emphysema is damage to the trachea
    b) Emphysema is damage to the alveoli
    c) Emphysema is damage to the heart
    d) Emphysema is damage to blood vessels

43. These muscles are directly involved in breathing:
    a) Pectoral muscles
    b) External obliques
    c) Intercostals
    d) Internal obliques

44. Bile:
    a) Aids protein breakdown
    b) Aids carbohydrate breakdown
    c) Aids nucleic acid breakdown
    d) Aids fat breakdown

45. Which hormone pair is involved in stressful situations?
    a) Thyroxin and calcitonin
    b) Adrenaline and noradrenaline
    c) Melatonin and histamine
    d) Glucagon and insulin

46. Which of these hormones is produced by the posterior pituitary gland?
    a) Cortisol
    b) Adrenaline
    c) Oxytocin
    d) Thyroid stimulating hormone

47. Vasopressin:
    a) Conserves body water (antidiuretic effect)
    b) Regulates blood glucose
    c) Stimulates uterine contraction during labour
    d) Raises metabolic rate

48. Inflammation of the gums is called:
    a) Diverticulitis
    b) Proctitis
    c) Gingivitis
    d) Enteritis

49. Which hormone is associated with sleep?
    a) Cortisol
    b) Melatonin
    c) Oxytocin
    d) Prolactin

50. Angina pectoris:
   a) Usually presents in puberty
   b) Commonly involves the heart
   c) Involves growth of endometrial tissue outside the uterus
   d) Is caused by obesity

Word Search 4 (Horizontal & Vertical)

F T H R O M B O S I S T E N D O N
U R T I C A R I A N V N H E N D P
X S F C X Y B S T R E S S W U G Y
K P K H O U S P O N D Y L O S I S
S R V L A L G S R F A S C I A F V
D A A O Z C S Y P H I L L I S L A
C I K A V E R R U C A V I P I E R
I N V S N R G T H R U S H C T P I
T Y E M O U U I T I N N I T U S C
Y X R A P W L I G A M E N T O M O
V I T I L I G O B N S T R A I N S
O T I T U B E R C U L O S I S X E
G R G S P O N D Y L I T I S H O Z
N B O X D U E S Y P H I L I S K F
N G B Q X Q W A R T T R V R O E A
J T O N S I L L I T I S V Y Z D X
O S T R O K E F T E T A N U S H C

ligament	tendon	fascia	vitiligo	chloasma	wart
verruca	vertigo	varicose	urticaria	ulcer	tuberculosis
tonsillitis	tinnitus	thrush	thrombosis	tetanus	stroke
strain	stress	syphilis	sprain	spondylitis	spondylosis

Anatomical Terminology

Fill the gaps using these words:

Flexion; Extension; Abduction; Adduction; Internal rotation; External rotation; Anterior; Posterior; Medial; Cranial; Lateral; Proximal; Distal; Coronal (frontal) plane; Transverse plane; Midline; Elevation; Depression; Pronation; Supination; Inversion; Eversion; Plantarflexion; Dorsiflexion; Circumduction; Atom; Molecule; Metabolism; Anabolism; Catabolism; Histology; Differentiation; Reproduction; Sagittal plane; Anatomical position; Ipsilateral; Contralateral; Caudal

1. In ________________ the angle at a joint is decreased.
2. ________________ turns the palm upwards.
3. During ________________ the foot twists to turn the sole inward.
4. Bending the foot in the direction of the plantar surface (sole) is called ________________________.
5. In ________________ the angle at a joint is increased.
6. A joint movement in which a distal part of the body moves in a circle is called ________________________.
7. In ________________ a body part is moved away from the midline of the body.
8. ________________ refers to a movement in which a body part faces downwards.
9. A ________________ ________________ divides an organ into superior and inferior sections.
10. A ________________ ________________ extends from side to side and divides an organ into anterior and posterior portions.
11. A ________________ ________________ divides an organ into left and right halves.
12. The ________________ is an imaginary vertical line that divides the body into equal right and left sides.
13. Bending the foot in the direction of the dorsum (upper surface) is called ________________________.
14. ________________ is nearer to, or at the front of the body.
15. ________________ is nearer to, or at the back of the body.
16. Nearer to the midline of the body or a structure is ________________________.

17. ________________ means towards or nearest the trunk of the body.
18. Away from the head or towards the lower part of a structure is ________________.
19. Standing with the legs together, feet flat on the floor, hands at the side and the arms facing forward is called the ____________________ ________________.
20. During ________________ the foot twists to turn the sole outward.
21. An upward movement of a part of the body is called ________________.
22. ________________ means away from, or farthest from the trunk of the body.
23. Farther from the midline of the body or a structure is ________________.
24. In ______________ ________________ the anterior surface of a bone or extremity moves towards the midline.
25. A downward movement of a part of the body is called ________________.
26. In ________________ a body part is moved towards the midline of the body.
27. Towards the head or towards the upper part of a structure is ______________.
28. In ______________ ________________ the anterior surface of a bone or extremity moves away from the midline.
29. An ___________ is the smallest constituent of an element that can take part in a chemical reaction.
30. ______________ is the set of chemical reactions that occur in living organisms in order to maintain life. These processes allow organisms to grow and reproduce, maintain their structures, and respond to their environments.
31. ________________ is the breakdown in living organisms of more complex substances into simpler ones together with a release of energy.
32. ______________ is the building of proteins and muscle mass in the body.
33. A ________________ is defined as at least two atoms in a definite arrangement held together by very strong chemical bonds.
34. The structural adaptation of some body part for a particular function is called ____________________. This occurs in the developing embryo.

35. ____________________ is the biological process by which new individual organisms are produced and is a fundamental feature of all known life.
36. The branch of biology that studies the microscopic structure of animal or plant tissues is called ____________________.
37. On the same side is called ____________________.
38. In the skull region is ____________________.
39. On the opposite side is called ____________________.

Histology

Fill the gaps using these words:

Cell membrane; Nuclear membrane; Nucleus; Nucleolus; Cytoplasm; Golgi apparatus; Mitochondria; Lysosomes; Endoplasmic reticulum; Ribosome; Centrosome; Centromere; Vacuoles; Centrioles; Chromatids; Interphase; Telomeres; Adipose tissue; Fascia; Ligaments; Tendons; Neuroglia; Cartilage

1. They connect bones together: ________________
2. Protects the contents of the cell and allows substances in and out: ____________________
3. Round granular structure within the nucleus of a cell, composed of protein and RNA, and functions primarily for the creation of ribosomes: ________________
4. The jelly-like substance that holds the organelles in place within the cell: ____________________
5. The 'brain' inside the cell, controlling all its activities including reproduction: ________________
6. They connect muscles to bones: ________________
7. A network of membranes found throughout the cell and connected to the nucleus, which functions as a manufacturing and packaging system: ____________________________________
8. Membrane surrounding the nucleus: ________________________
9. Organelle responsible for packaging proteins into little bubble-like sacs called vesicles and sending them to the correct destinations within the cell: ______________________________
10. Rod-like organelles that appear when a cell is about to divide: ____________________
11. Resting phase of cell division: __________________
12. Chromatin strands of a copied chromosome: ________________
13. Supports the neurons in the nervous system: ________________
14. Stable chromosome ends that protect the DNA: ________________
15. Storage bubbles inside cells holding nutrients as well as waste: _________________
16. Also called fatty tissue: ______________________
17. An interwoven system of connective tissue throughout the body: _________________

18. A strong but flexible tissue composed of cells called chondrocytes: ____________________

19. A darker, more condensed area of the cytoplasm, which appears when a cell is about to replicate: ________________________

20. Regarded as the energy generators within cells: ____________________

21. Sometimes referred to as the waste disposal units within cells: ______________________

22. Organelles that build protein within cells - they connect one amino acid at a time and build long protein chains: ___________________

23. The part of the chromosome in the middle that appears pinched together: ___________________

Skin/Scalp Disorders

Fill the gaps using these words:

Acne vulgaris; Allergic reaction; Albinism; Anhidrosis; Alopecia areata; Basal cell carcinoma; Boils; Broken capillaries; Bromhidrosis (osmidrosis); Burns; Candida; Cellulitis; Chloasma; Comedones; Conjunctivitis; Crow's feet; Eczema/dermatitis; Ephilides; Folliculitis; Lice; Herpes simplex; Herpes zoster; Hyperhidrosis; Impetigo; Keloid; Lentigo; Malignant melanoma; Milia; Milaria rubra; Moles; MRSA; Naevae; Port wine stain; Pressure sores; Psoriasis; Scabies; Scleroderma; Seborrheic dermatitis; Keratosis pilaris; Squamous cell carcinoma; Stye; Systemic lupus erythematosus (SLE); Striae; Tinea capitis; Tinea corporis; Tinea pedis; Urticaria; UV damage; Vitiligo; Warts/Verrucas; Verrucae filliformis

Bacterial skin diseases: pathogenic bacteria cause disease in two ways; they gain access to the tissues and multiply to damage their surroundings or they release toxins, which poison remote parts of the body. Antibiotics are commonly used to kill or inhibit the multiplication of bacteria.

1. ____________________________________: This non-contagious condition is often encountered in adolescence and pre-menstrually, but can proceed into adult life. It is an inflammation of the sebaceous glands in the skin and is normally caused by hormonal imbalances that increase sebum production leading to blocked glands and infection. The skin has a shiny, sallow appearance with pustules and comedones. It is prone to open pores. When pustules clear, they may leave pitting and scarring of the skin. The main sites of infection are the face, back, chest and shoulders.
2. _________________________________: A bacterial infection causing thin roofed blisters that weep and leave a thick, yellow crust. Can occur at any age but more common in newborns and in children. It causes itching and is highly contagious.
3. ____________________: This is a painful red swelling in the skin (also known as a 'furuncle') caused by bacterial infection around a hair follicle. Several may join to form a 'carbuncle'.

4. ______________________: This is a bacterial infection of the pilo-sebaceous duct (sebaceous gland and hair follicle) causing inflammation. It is common in adolescence and often linked with acne vulgaris. It is called 'sycosis barbae' in the beard area.
5. ______________________________: Inflammation of the mucous membrane that covers the front of the eye and lines the inside of the eyelid. It becomes red and swollen and produces a watery or pus-containing discharge. It is highly contagious and can be caused by bacteria, virus, allergy, or physical or chemical irritation.
6. ____________: Methicillin-resistant staphylococcus aureus, a bacterium with antibiotic resistance. Infection mainly occurs in those who are already ill and hospitalised. It can be difficult to treat as the bacteria are resistant to most types of antibiotics. Many people are carriers without realising it, as it often does not cause any symptoms in healthy people. If it invades the skin through a cut, it can cause skin infections such as boils, pimples, impetigo and skin abscesses, and is a common cause of wound infections. In some people, it can get into the bloodstream and travel to internal parts of the body to cause more serious infections such as sepsis. It is spread through contact with infected individuals and clothing.
7. __________________: A spreading bacterial infection of the skin that begins as redness and tenderness and develops into hot swollen areas that are slightly pitted. It can occur after being bitten by dogs or humans, or an injury in dirt or water.
8. ____________: Also called a hordeolum, this is a bacterial infection of the sebaceous glands at base of the eyelashes. It lasts 2-4 days and is characterised by inflammation, tenderness, pain and later a pus-filled cyst.

Viral infections: viruses are resistant to antibiotics; prevention of disease is by avoidance of contact with infected individuals or by vaccination.

9. ____________: Small solid benign tumours that arise from the surface of the skin as a result of proliferation of the papilloma virus in skin cells. They are usually found on fingers and thumbs. Those found on the feet are called ____________ and both are contagious.

10. ____________________: Also called 'cold sores', these are caused by a virus which most people carry without symptoms. They are not confined to the mouth and can spread to other parts of the face and body. They appear as small blisters, which if left alone will dry up leaving a crust that falls off. They are highly contagious when active.
11. ____________________: Commonly called 'shingles', this is also known as 'Zona', a Greek word meaning 'belt' or 'sash' due to the fact that the accompanying rash often appears in a band around the torso. It is the adult form of chickenpox. It most commonly attacks adults (who have had chickenpox previously), often after contact with a child suffering from chickenpox, and usually when the person is stressed or physically run down. The affected areas are always along the course of one or more of the spinal or cranial nerves beneath the skin, and it is highly contagious when the rash is in the blister phase.

Fungal infections: develop from living organisms, which grow and multiply on the body. Fungal infections usually itch.

12. ______________________________: Also known as body ringworm, this highly contagious fungal infection usually affects children and is normally caught from pets or animals. It presents as an itchy reddened area with raised edges. The classical presentation is that of a circular, scaly lesion that is clear centrally and slowly spreads like ripples, thus creating the appearance of a ring. It affects the trunk, face or limbs and frequently begins in body folds such as the groin or axilla. It takes about 2 weeks for symptoms to develop, and can be transmitted by a person who is carrying the fungus without any symptoms being present.
13. ____________________: Also known as 'athletes foot', this contagious condition is common in adolescents and young adults who wear heavy footwear (increased sweating) and those who use communal changing facilities, showers and swimming pools. It usually starts in the toe webs as a patch of sodden, irritable skin with redness and scaling, and in some cases there may be acute pustular eruptions. Sometimes a dry scaling rash extends across the sole and dorsal surface of one foot and this can be confused with psoriasis or eczema. It is so common that it affects about half the population at one time or another.

14. ______________________________: Also known as scalp ringworm, this condition is found mainly in children under ten and is virtually never present after adolescence. There are a number of round scaly patches on the scalp from which the hair falls out. The characteristic signs of this condition are circular bald patches with redness, scaling, and brittle hair stumps in the patches. It is usually caught from contact with cattle or other domestic animals.

15. __________________: This is a type of yeast that lives in the body. When the immune system is working properly, the amount of this yeast in the body is kept in balance. When the immune system is compromised or there are other factors such as consuming unhealthy amounts of sugars and carbohydrates, the yeast can grow out of control infecting the mouth, skin, intestines or vagina.

Infostations: caused by organisms called parasites that live either on the body (lice) or in the body (tapeworms).

16. _________________: A highly contagious skin disease caused by infestation with a parasitic 'itch mite'. The disease is spread by close contact with infested persons. The female of the parasite burrows into an area of skin e.g. on the front of the wrist, the sides and webs of the fingers, buttocks, genitals or feet. It does not occur above the neck. Eggs are deposited in the small tunnels she makes, while the male remains on the surface of the skin. The burrowing activities of the parasites cause intense itching, and the original small blister-like lesions are made worse by scratching. Treatment works best if all of the patient's close contacts and family are treated for the disease. Itching may persist even after all the mites have been eliminated. Secondary infection as a result of scratching may need to be treated.

17. ____________: (Pediculosis capitis/corporis/pubis) affects the scalp, though sometimes involves the eyebrows, eyelashes and beard. It can also affect the whole body or just the pubic region in adults. Scalp infestation is particularly common in children, especially girls. These parasites feed on blood from the scalp and are transmitted by direct contact with hair or with items such as combs, towels and headgear. The bites cause severe persistent itching, and the lesions may become infected. They may be seen particularly around the back of the head and behind the ears. The small ovoid eggs are easier to detect as they are firmly attached to the hair shafts. These hatch in 3 to 14 days. A fine comb may remove them, and the scalp should be treated. Members of the same household should also be examined for infestation.

Pigmentation Disorders: due to variations in melanin or other causes.

18. __________________: The partial or complete loss of melanocytes from patches of skin leading to irregular white areas. It is thought to be an autoimmune disease in which the body produces antibodies that destroy melanocytes. The areas affected tend to gradually increase and there may be other autoimmune diseases in the same patient. A family history of the condition is found in one third of patients. It can affect skin of any colour but is more obvious in darker skins. Leoucoderma (Leukoderma) is another name for this condition.
19. _________________: An inability of an individual of any race to produce melanin, which is the pigment in the skin that varies skin colour from pale yellow to black. This defect is hereditary, but is caused by a recessive gene - which means that it will appear only when two people who both carry the gene have children. Even then, it occurs in only about one in four of their children. Sufferers have poor eyesight and extreme ultraviolet sensitivity.
20. _________________: Localised brown colouration (butterfly mask) in the skin of the face - particularly on the cheeks - associated with pregnancy, the contraceptive pill or menopause. The cause is thought to be a disturbance in hormone production, which leads to a rise in the amount of melanin formed in the deeper layers of the skin. It usually disappears spontaneously at the end of pregnancy, when the pill is stopped, or after the menopause.

21. ___________________: Commonly known as 'freckles', these are small, pigmented areas of skin, not usually raised and less than 0.5 cm in diameter. They occur in areas exposed to the sun in fair skinned people and fade during the winter. They are harmless and are regarded by some as a sign of beauty.

22. ______________________: This occurs characteristically in areas exposed to the sun in elderly people. Also known as liver spots, these dark patches of pigmentation appear more distinct than freckles and have a slightly raised appearance and more scattered distribution.

23. ______________: A common occurrence on the face and body and present in several different forms, varying in size, colour and vascular appearance. They are usually slightly raised and dark, and sometimes hairy. Flat ones are called 'sessile', while those raised above the surface or attached by a stalk are referred to as 'pedunculated'.

24. _______________: Clearly defined skin markings that are present at birth. Melanocytic (from cells that produce melanin) and vascular (from blood vessels) are generally the most common types of birthmarks. Pigmentation varies from light brown to red to black, depending on the type. The 'strawberry' birthmark affects babies, usually disappearing after a few years.

25. ___________________________: Vascular birthmark due to large area of dilated capillaries causing a pink to dark red skin colour that makes it contrast vividly with the surrounding skin - commonly found on the face. These birthmarks last into adulthood and the colour usually gets darker, turning purple or a deep red.

Abnormal skin growth: due to various causes.

26. ____________: Results from a greatly increased turnover of epidermal cells. The normal period from new cell to desquamation is 28 days, but for sufferers of this condition, this is reduced to around 4 days. This is a chronic inflammatory skin disease characterised by red patches covered with silvery scales that are constantly shed. Sizes of scales vary from minute spots to quite large sheets of skin. Points of bleeding may occur beneath the scales. It can affect whole body or a specific area like the scalp or face and is not contagious.

27. ______________: An overgrowth of scar tissue over a healed wound. It is usually smooth, shiny and flesh coloured. It is not painful but can itch and be sensitive to touch. It can happen to anyone, but is more common in people with darker skin.
28. ____________________________: Senile warts, which are common in elderly people. They are harmless, flesh coloured, brown or black growths - common on the torso and temples but can appear anywhere.
29. ____________________________: Also known as a 'skin tag', this is a long, slender wart that is common on the thin skin of the eyelids, armpits or neck, and often disappears spontaneously.

Allergies: inappropriate response to normally harmless substances - usually cause redness, inflammation, swelling and pain.

30. ____________________________: When irritated, the body produces histamine, which is part of the skin's defense mechanism. This can cause red blotchy patches on the skin as well as watery, stinging eyes, swelling and a runny nose. Symptoms can be slight or intense, depending on individual reactions.
31. __________________: This is also called 'hives' or 'nettle rash' and is often an allergic reaction. It is characterised by wheals or welts of pinkish colour produced by extreme dilation of capillaries. It is very itchy and can lead to secondary bacterial infection by vigorous scratching that tears the skin.
32. ________________________________: Both terms are often used synonymously, although the latter is usually caused by contact with external irritants. The condition can be found all over the body but most often on the inside of the knee and elbow joints, and on the face, hands and scalp. The skin becomes very dry and itchy causing great discomfort. There can be dry, scaly patches with bleeding at points, but it is not contagious.

Skin Cancers (carcinomas): the development and later spread throughout the body of cells from a malignant tumour. Possible causes are (1) Radiation: UV radiation is the most common cause of epidermal malignancy. (2) Chemicals: some chemicals are associated with malignancy. (3) Immune responses (4) Genetic

33. ________________________: This malignant tumour - which is the most common of the three - accounts for 75% of skin cancers and most commonly affects the face. It presents as a firm nodule, clearly growing within the skin and below it rather than on the surface. The tumour is composed of masses of dividing basal cells that have lost their capacity to differentiate any further. As a result, no epidermis is formed over the tumour and the surface breaks down to form an ulcer, sometimes referred to as a 'rodent ulcer'. The residual edges of the nodule form a characteristic 'rolled edge'. Once the basal cells have invaded the deeper tissues, the rolled edge disappears. If untreated it will slowly erode underlying structures. It is easily treated when small, and rarely metastasises.

34. ___________________________: This affects cells found on the surface of the body, on the top layer of the skin, and can be caused by sunlight, chemicals or physical irritants. It starts very small but grows rapidly, becoming raised. Most are easily removed by laser or by cutting away in a minor operation. When the carcinoma is large a skin graft may be needed to cover the damage.

35. ___________________________: This is the most serious form of skin cancer because unlike the others, it can spread quickly throughout the body. It affects the skin cells that produce melanin. Often, a mole, which has been present since childhood, changes - it may begin to grow, become patchy, lighter or darker, develop a black margin that spreads into the surrounding skin, bleed spontaneously or itch. Another symptom is the development of a new mole or a flesh-coloured or red lump may appear on the skin. It may also develop in the dark, irregular freckles that sometimes occur on the skin of elderly people. Most tend to arise in areas exposed to the sun, but they can occur on other sites such as the soles of the feet or under the nails.

Other disorders: various causes (sebaceous gland or sweat gland abnormalities, autoimmune conditions etc.)

36. _________________: Also called hypohidrosis, this is a lack of sweating in the presence of appropriate stimuli such as heat. It can be congenital or due to illness.

37. ______________________________: Body odour caused by breakdown of sweat by bacteria and yeasts that normally live on skin. It can be influenced by diet, genetics and general health.
38. ___________________: Excessive or constant sweating, affecting the entire skin or limited to certain areas such as the armpits, genitals, palms or soles. It can be due to illness, hyperthyroidism or certain drugs.
39. ____________________________: Also called prickly heat, this condition is due to trapped sweat and is common in warm, humid climates. The sweat gets trapped in the sweat ducts and causes inflammation and a prickling, itching sensation. It can appear as an itchy rash of small blisters or large red areas of skin.
40. ________________________________: This common skin condition looks like small goose bumps, but is actually dead skin cells that build up in sebaceous glands around hair follicles. It usually affects the upper arms and/or legs.
41. _____________________: This is a common skin condition that mainly affects the scalp. It causes scaly patches, red skin and stubborn dandruff. It can also affect oily areas of the body such as the face, sides of the nose, eyebrows, ears, eyelids and chest.
42. ____________________________: An autoimmune condition, also known as spot baldness. Often it results in a few bald spots on the scalp, each about the size of a coin. Psychological stress may result, but sufferers are generally otherwise healthy.
43. ____________: Also called 'stretch marks', this is a form of scarring under the skin with an off-colour hue. They are due to tears in the collagen and elastin fibres of the dermis, and over time can diminish but not disappear completely. They often result from stretching of the skin associated with rapid growth (common in puberty) or weight gain (e.g. pregnancy or muscle building) that overcomes the dermis's elasticity.
44. _____________________: These are fine lines around the eyes caused by habitual expressions and daily movements. They are associated with ageing muscle tissue, but premature formation may be due to eyestrain and is often associated with oedema (swelling) around and under the eyes.

45. ______________________________: These look a bit like red spiders and can appear on the thighs, nose and cheeks as a person gets older. The condition affects women in particular and some more than others. Capillary walls are normally very elastic, and dilate when the skin is hot or as a response to spicy foods, alcohol or other stimuli. When this happens the person looks flushed, but as soon as the stimulus wears off the capillaries should return to normal. If the capillary walls lose their elasticity due to poor diet, excessive alcohol, excessive intake of stimulants such as strong tea or coffee, very severe climates, or circulatory disorders, the capillaries may not shrink as they should and the face can appear red all the time.

46. ______________________: Sunlight is important to health, particularly because the skin manufactures Vitamin D when exposed to sunlight. As little as ten minutes in the open air each day, even in winter, is good for the health. However, over exposure to the ultra violet rays of the sun weakens the bonds between cells in the stratum corneum, which is the top layer of skin. The stratum corneum is the body's first line of defense, and it is critical to apply sunscreen daily and throughout every season to prevent premature skin ageing or skin cancer.

47. ______________________: Commonly known as 'blackheads', these are caused by a build-up of sebaceous secretions that have become trapped in hair follicles and have subsequently dried out and hardened. The colour comes from oxidation rather than dirt, and they are common in puberty.

48. ______________________: Also known as 'white heads', these form when sebum becomes trapped in a blind duct with no surface opening. The condition is most common on dry skin and can appear on the orbicularis oris muscle around the eyes, or between the eyebrows. They can also appear as small keratin cysts on the cheeks and eyelids.

49. ________________________: Also known as bedsores, this is a type of injury that affects areas of the skin and underlying tissue. They most often develop on skin that covers bony areas of the body, such as the heels, ankles, hips and tailbone. They are caused when the affected area of skin is placed under excessive pressure. It can range in severity from patches of discoloured skin to open wounds that expose the underlying bone or muscle. Bedsores can be a serious problem among frail older adults, and can be related to the quality of care the person receives - they need to be turned, positioned correctly, and given good nutrition and skin care. People with diabetes, circulation problems, and poor nutrition are at higher risk.

50. __________________: This is a connective tissue disease that involves thickening and hardening in the skin and can also affect blood vessels, muscles, and internal organs. It is a type of autoimmune disorder - a condition that occurs when the immune system mistakenly attacks and destroys healthy body tissue. The cause is unknown but people with this condition have a build up of collagen in the skin and other organs.

51. ____: This is a chronic autoimmune disease that causes inflammation in various parts of the body. It is commonly called 'lupus'. The severity ranges from mild to severe. There are two main forms of lupus. Discoid lupus only affects the skin while the other form affects the skin and joints, and may also involve internal organs such as the heart or kidneys. The symptoms and severity of this condition vary tremendously between people. Many people have fatigue, weight loss and a mild fever. In addition, it can cause joint and muscle pain, skin rash, hair loss, sensitivity to sunlight, poor circulation in fingers and toes, mouth ulcers, anaemia, anxiety and depression. Steroids and immune suppressant drugs are usually prescribed as treatment.

52. ______________: This is skin damage caused by contact with fire, heat, electricity, radiation, or caustic chemicals. Damage is classified according to the depth and extent of the skin damage (first, second and third degree).

53. __________________: These are small itchy painful lumps that develop on the skin as an abnormal response to cold.

Nail Diseases (Onychosis)

Fill the gaps with these words:

Ridges/furrows; Beau's lines; Pitting; Leukonychia; Onychorrhexis; Hangnail; Blue nails; Black nails; Stained nails; Melanoma; Onycholysis; Onychauxis; Onychogryphosis; Onychatrophia; Koilonychia; Onychocryptosis; Onychophagia; Psoriasis; Dermatitis/Eczema; Tinea ungium; Paronychia; Whitlow; Corns; Onychia; Tinea pedis

1. __________________: A bacterial infection of the cuticle.
2. __________________: Painful thickenings of the skin caused by excessive pressure or friction.
3. ______________________: Can be the result of age or ill health as well as trauma or constant picking at the nails.
4. __________________: A sign of an underlying problem such as dermatitis or psoriasis.
5. ____________________: Presents as white spots on the nails generally caused by injury to the nail matrix allowing an air pocket to form under the nail.
6. ______________________: Horizontal ridges or grooves on the nail plate and can reflect temporary retardation of growth due to illness, high fever or nutritional deficiency e.g. zinc.
7. ____________________: A piece of skin that appears jagged and torn and generally appears on the fingers and not on the toes, though it's possible to have one around a toenail.
8. ______________________: Dry, brittle, splitting nails usually due to reduced blood supply or contact with strong detergents.
9. ___________________________: Can be due to nicotine, dark nail enamels or hair dyes.
10. ___________________________: Usually due to heavy bruising.
11. __________________________: May be due to poor circulation, anaemia or a heart problem.
12. __________________: Separation of the nail plate from the nail bed and can be due to an overactive thyroid, fungal infection, psoriasis, iron deficiency, trauma or infection.
13. ____________________: A form of skin cancer that can affect the nails.

14. ________________________: Atrophy of the nail, which becomes smaller and sometimes sheds entirely.
15. _________________________: An enlarged, thickened nail, with an increased curve (ram's horn nail) and is often due to neglect.
16. _________________________: An ingrown nail.
17. _________________________ Spoon shaped nails - can be hereditary or due to iron deficiency anaemia or overactive thyroid.
18. __________________________: Technical term for bitten nails.
19. ______________________________: Often results from contact with irritants or can be genetic.
20. _________________: A painful infection of the skin around the nail, usually due to the herpes simplex virus.
21. ___________________________: Ringworm of the nail.
22. ___________________________: Also known as athlete's foot.
23. ______________: Characterised by raw scaly skin and can be confused with eczema. When it attacks the nail plate, it usually leaves it pitted and dry.
24. ______________: Inflammation of the nail bed, with the development of pus and loss of the fingernail or toenail.
25. ______________: An abnormal thickening of the nail plate on fingers or toes.

Skeletal System Disorders

Fill the gaps with these words:

Arthritis; Ankylosing spondylitis; Bunions; Bursitis; Carpal tunnel syndrome; Cervical spondylosis; Fracture; Ganglion; Gout; Kyphosis; Lordosis; Osteo-arthritis; Osteoporosis; Prolapsed (herniated / slipped disc); Rheumatoid arthritis; Scoliosis; Whiplash; Prostheses; Osteomalacia; Osteogenisis imperfecta; Psoriatic arthritis; Paget's disease; Spinal stenosis

1. ________________________________: Degeneration of intervertebral discs and neck vertebrae, usually affecting older people. The area can become inflamed leading to spondylitis. It causes neck pain and pressure on the spinal cord.
2. _______________: Painful swelling of the joint between the big toe and first metatarsal. Bursitis can develop and big toe can become laterally displaced (hallux valgus). It can be hereditary or be caused by ill-fitting shoes.
3. _______________: Inflammation of a bursa due to overuse or irritation from unusual use. It causes pain, inflammation and limited movement and may be due to injury, gout or arthritis.
4. _______________: A type of cyst, also known as a 'bible bump' from a common urban legend that treatment by pounding on the cyst with a bible or another large book occurred in the past. It is a benign swelling that often appears on or around joints and tendons in the hand or foot.
5. _________________: Artificial extensions that replace missing body parts.
6. _____________: A painful condition caused by the buildup of uric acid crystals in the tissues, especially the big toe, but it can affect other areas.
7. _____________________: Chronic degenerative arthritis of joint cartilage leading to pain and stiffness. It is caused by overuse, wear and tear. It is common in weight-bearing joints and is often associated with ageing. Affected joints can often be surgically replaced e.g. hip or knee replacement.
8. ___________________: An injury to the neck (the cervical vertebrae) resulting from rapid acceleration or deceleration - as in a car accident.

9. ____________________: Porosity and brittle bones common in post-menopausal women. Caused by a number of factors including decreased hormone production (oestrogen), calcitonin /parathormone imbalance, decreased calcium or Vitamin D intake and lack of weight-bearing exercise, especially during bone deposition years - up to about 26 years old.
10. ____________________________________: Caused by increased pressure on intervertebral discs, resulting in bulging or rupturing.
11. ___________________________: An autoimmune condition affecting cartilage and joint lining. The synovial membrane inflames, thickens and fluid accumulates, causing pain and tenderness. The membrane produces granulation tissue that erodes cartilage causing bone ends to fuse. It affects smaller joints first and is always bilateral. It can be hereditary and can affect any age group.
12. ______________________: A general term for inflammation of the joints but there are varying types of this disease.
13. ________________: An increased inward curvature of the lumbar spine.
14. ______________________________: A chronic inflammatory disease of the spine and nearby joints causing pain and stiffness in the spine, neck, hips (sacroiliac joint), jaw, and rib cage. If ribs are affected, it can be difficult to breathe deeply. It mainly affects men between 20-40 and there is a genetic link. It causes back pain, stiffness, muscle spasms, weight loss, fatigue and other symptoms such as inflammation in the eyes.
15. ________________: Increased outward curvature of thoracic spine (hunchback).
16. ___________________________: Also called brittle bone disease, this is a genetic disease causing brittle bones. Symptoms include frequent fractures, bone deformity, discolouration of sclera of the eyes, deafness and thin dental enamel.
17. ____________________________________: A painful disorder caused by compression of the median nerve in the wrist. It causes discomfort and weakness in the hands and fingers and sensations of tingling, burning or numbness. It can occur during pregnancy due to an accumulation of fluid in the tissues.

18. ________________________________: Inflammation of the joints, usually in people who already have psoriasis. Sometimes the condition develops before the psoriasis and sometimes affects one or two joints e.g. knee or ankle, but often several joints, both large and small and on both sides of the body, are involved. About a third of sufferers also have a painful, stiff back or neck caused by inflammation in the spine (spondylitis). It can affect the nails with pitting, discoloration and thickening. Joints at the ends of the fingers or toes (distal inter-phalangeal joints) can also be inflamed. This condition is more common in adults than children.

19. ________________________: A condition characterised by excessive overgrowth of bone, especially in the spine, pelvis, skull, and femur. There are often no symptoms but pain, deformity and fracture can occur.

20. ________________________: Softening of the bones caused by a deficiency of vitamin D. This condition is also called 'rickets' in children.

21. ____________________________: A medical condition in which the spinal canal narrows and compresses the spinal cord and nerves.

22. ______________ A break of bone due to injury or disease; there are several types.

23. __________________: This is a sideways curvature of the spine.

Muscular System disorders

Fill the gaps with these words:

Atony; Atrophy; Adhesive capsulitis; Adhesions; Bursitis; Cramp; Fibromyalgia; Golfer's elbow; Housemaid's knee; Lumbago; Myositis; Microtrauma; Muscle fatigue; Muscular dystrophy; Myasthenia gravis; Plantar fasciitis; Repetitive strain injury (RSI); Rheumatism; Shin splints; Spasm (X2); Sprain; Strain; Spasticity; Tendonitis; Tetanus; Tennis elbow

1. ______________: Caused by lack of nourishment or use and resulting in wasting of muscle or decrease in size.
2. ______________: Inflammation of a tiny fluid-filled sac that functions as a gliding surface to reduce friction at a joint in the body.
3. _______________: Can be due to overexertion, extreme heat or sodium / water deficiency, causing involuntary, localised painful contraction of one or more muscles.
4. ______________________: Widespread pain in the fibrous tissues of the body (muscles, tendons, ligaments). It is often associated with chronic fatigue and stress and is more common in women than it is in men.
5. ____________________________: (AKA frozen shoulder) is a painful condition which affects the ability to move the shoulder, and usually only occurs on one side.
6. ______________________: The term often given to any pain at the front of the lower leg. Symptoms occur at the front inside of the shinbone and can arise from a number of causes e.g. running.
7. ______________: Lack of normal tone or tension in muscle.
8. ______________: Fibrous growths similar to scar tissue that form at the site of tissue damage.
9. _______________: Injury or damage to a joint caused by an acute episode such as a sudden twist, resulting in pain and swelling of the ligaments.
10. ________________: Overexertion or overstretching of a muscle that may result in bleeding or possibly tearing of the muscle, fascia or tendon.

11. ________________: Inflammation in or around tendons - bands of strong fibrous tissue that hold muscle to bone. Inflammation of the Achilles tendon in the heel is common in runners. In some cases there is no inflammation but pain due to wear and tear. It can be due to tight hamstrings or high arches pulling on the tendon.
12. ________________: An acute infectious disease usually contracted through a penetrating wound. It causes severe muscular spasms and rigidity in the jaw and neck and if not treated these can eventually affect the whole body. It is caused by the bacterium 'clostridium tetani' found in soil and the intestines and faeces of animals.
13. ________________: The general term for inflammation of muscle.
14. ________________: Small but widespread tears in muscle cells from training stress.
15. ________________________: An extremely common and painful injury appearing in a large proportion of tennis players. It occurs most commonly in the tendon of the extensor carpi radialis brevis muscle, approximately 2cm below the outer (lateral) edge of the elbow joint (at the lateral epicondyle).
16. ________________________: Can be a result of vigorous exercise but abnormal fatigue may be due to excessive high intensity training and diminished rest.
17. __________________: A non-specific term for medical problems affecting the joints and connective tissue. The study of, and therapeutic interventions in such disorders is called rheumatology.
18. ______________________________: Usually associated with injuries to the forearm, wrist and hand but can occur throughout the body - caused by repetitive movements of the wrist or fingers.
19. ____________________________: A similar injury to tennis elbow but causes pain on the inside (medial side) of the elbow instead (at the medial epicondyle).
20. ____________________________: Inflammation of the pre-patellar bursa, which lies in front of the patella (kneecap).
21. ________________: Backache affecting the lumbar region (lower back) and can be caused by muscle strain, arthritis, or a ruptured intervertebral disc.

22. ________________________: A group of inherited disorders, which destroy muscles leading to muscle weakness. They are generally characterised by a progressive atrophy of skeletal muscles due to degeneration of individual muscle fibres. There are several types including Duchenne, caused by lack of a protein called dystrophin. About 100 boys with Duchenne are born in the UK each year.
23. ______________: A CNS disorder in which certain muscles receive continuous messages to tighten and contract. This causes stiffness and rigidity of muscles and interferes with movement and sometimes speech.
24. ______________________________: An autoimmune disease causing weakness of skeletal muscles due to impaired communication between nerves and muscles. It is more common in women and symptoms include drooping eyelids, weak eye muscles, double vision, extreme fatigue, difficulty speaking and swallowing and weakness of arms and legs. There are usually remissions and relapses in the condition.
25. ______________: An abnormal, involuntary contraction that may occur as part of another disorder or as a localised condition. A cramp is a type of ____________ as is a tic (twitching of eyelid or facial muscles).
26. ______________________________: An accumulation of micro-trauma to the plantar fascia – the thick fibrous connective tissue extending from the heel bone to the toes. It causes pain in the underside of the heel and difficulty dorsiflexing the foot. Stretching the fascia and calf muscle helps with pain.

Cardiovascular System Disorders

Fill the gaps with these words:

Angina; Anaemia; Aneurysm; Embolus; Gangrene; Haemophilia; Haematoma; Hemorrhoids; Atherosclerosis; Arteriosclerosis; Coronary thrombosis; High cholesterol; Hypertension; Hypotension; HIV & AIDS; Hepatitis; Palpitations; Pulmonary embolism; Raynaud's disease; Sickle cell anaemia; Thalassaemia; Varicose ulcers; Varicose veins; Cardiac arrhythmia / tachycardia / bradycardia; Epistaxis; Hole in the heart; Intermittent claudication; Myocardial infarction; Leukaemia; Phlebitis; Septicaemia; Thrombus

1. ________________: A reduction in the ability of the blood to carry oxygen, either by a decrease in red blood cell production or a failure of the haemoglobin in the red blood cells.
2. ________________________________: A vascular disorder that affects blood flow to the extremities (fingers, toes, nose and ears) when exposed to cold temperatures, or in response to psychological stress.
3. ________________________________: Irregular heartbeat; ________________________ abnormally rapid heartbeat (over 100 beats per minute); ________________________ a slow heartbeat, defined as fewer than 60 beats per minute for an average adult.
4. ______________: A recurring pain or discomfort in the chest that happens when some part of the heart does not receive enough blood.
5. _______________: A localised, blood-filled dilation (balloon-like bulge) of a blood vessel caused by disease or weakening of the vessel wall.
6. __________________: A large bruise or collection of blood under the skin, producing discolouration and swelling in the area; usually caused by trauma.
7. ________________________: Also called piles; these are swelling and inflammation of veins in the rectum and anus. The anatomical term technically refers to 'cushions of tissue filled with blood vessels at the junction of the rectum and the anus.'
8. _____________________: Degenerative disease causing hardening of the arteries with loss of elasticity and can lead to an increase in blood pressure.

9. ________________________: A buildup of fats e.g. cholesterol, inside the artery walls. Eventually leads to hardening of the arteries and an increase in blood pressure.

10. ____: Formation of a clot in one of the arteries that carries blood to the heart muscle – can lead to myocardial infarction (MI) commonly known as a heart attack, and the death of some cardiac muscle cells.

11. ______________________: High blood pressure of 140/90 or above.

12. ______________________: Low blood pressure with systolic pressure less than 90 millimetres of mercury (mm Hg) or diastolic less than 60 mm Hg.

13. ____________: Human Immunodeficiency Virus & ______________ Acquired Immune Deficiency Syndrome – a disorder of immune system (T-cell destruction) leaving the body vulnerable to infection. It is transmitted by blood and body fluids.

14. ________________________________: Chronic ulceration above the ankles due to varicose veins that interfere with normal blood circulation in the affected areas.

15. __________________: Blood clot within the vascular system – can obstruct blood flow in an artery (e.g. coronary thrombosis); can also occur in a vein (deep vein thrombosis) and may be associated with inflammation (thrombophlebitis). This may produce pain and local swelling. If a portion breaks free, it can be carried up to blood vessels in the lungs and may produce severe chest pain or cardiac arrest.

16. ________________________: An abnormal awareness of the heartbeat, whether it is too slow, too fast, irregular, or at its normal frequency.

17. ____: Blockage of a pulmonary artery by foreign matter or by a blood clot.

18. __________________: Material such as blood clot, fat or air bubble that is carried in the blood from one point of the circulation to another – can become lodged in an artery and obstruct blood flow.

19. ____________________: Inflammation of a vein, usually in the legs.

20. ________________: This means 'liver inflammation' and is caused by a virus. Several variations of the virus can infect the liver - the most common are A and B. A is transmitted through food and water contaminated with faeces due to poor personal hygiene or sanitation. B is spread through the exchange of blood and body fluids. C is also spread through the exchange of blood or blood products. It can be spread through the sharing of drug-use equipment or needle stick accidents (wounds caused by needles that accidentally puncture the skin). It was also spread by blood transfusions before September 1992, when screening was introduced.
21. __________________: An inherited disease that is due to a deficiency or lack of certain clotting compounds in the blood.
22. ____________________: An inherited form of anaemia caused by faulty synthesis of haemoglobin. It particularly affects those of Mediterranean, Middle Eastern and South Asian origin.
23. ____: A severe hereditary form of anaemia, mainly affecting people of African origin, in which mutated haemoglobin distorts red blood cells into a crescent shape, causing the cells to become stuck in capillaries.
24. __________________: A type of fat (lipid) made by the body that is essential for good health and found in every cell in the body. There are two different types – high density (HDL) and low density (LDL). Having a high level of LDL in the blood can increase the risk of heart disease and stroke by creating sticky deposits (plaques) along the artery walls and may lead to heart disease. This is primarily caused by diet and family history. HDL is regarded as 'good', because it carries LDL from other parts of the body back to the liver, which then removes it from the body.
25. ________________: Nosebleeds – the nose is rich in blood vessels and can bleed easily but if it occurs regularly it should be medically investigated.
26. ________________: The death of body tissues due to poor blood supply.
27. ____________________________: (Septal defect) - small hole or holes in the septum (wall) between the atria and ventricles of the heat.
28. _______________________: Twisted, widened veins caused by swollen or enlarged blood vessels, due to a weakening in the vein wall or valves.

29. ________________________________: Lameness due to pain in the leg muscles because the blood supply is inadequate - sometimes due to blocked arteries. The pain usually subsides with rest.

30. ________________________________: Also known as a heart attack, this refers to death of heart muscle tissue due to loss of blood supply (cardiac ischemia) – often due to a blood clot that blocks the flow of blood through a coronary artery. Heart attack victims are often warned of trouble by episodes of angina. This is chest pain that, like a heart attack, is provoked by ischemia. Most heart attacks are the result of coronary artery disease, also known as atherosclerosis or 'hardening of the arteries', a condition that clogs coronary arteries with fatty, calcified plaques over time.

31. ____________________: A type of cancer that affects the white blood cells which become abnormal, dividing and growing in an uncontrolled way. This crowds out the space for red blood cells and leads to anaemia and a weak immune system.

32. ____________________: Rapid multiplication of bacteria and the presence of bacterial toxins in the blood; commonly known as blood poisoning.

Lymphatic & Immune System Disorders

Fill the gaps with these words:

Allergies; Glandular fever; HIV/AIDS; Hodgkin's lymphoma; Non-Hodgkin's lymphoma; Leukaemia; Lymphoedema; Tonsillitis; Systemic lupus erythematosus (SLE)

1. ______________________________: Malignant lymphoma characterised by painless enlargement of one or more groups of lymph nodes. There may also be weight loss, fever, itching and profuse sweating at night. It slowly spreads to adjacent lymph nodes but rarely metastasises to distant sites. The cause is unknown but great progress has been made in the treatment of this disease, such that it is now thought to be curable in many cases.
2. ____________________________________: There are at least 20 different types of this condition representing a broad category of cancers that arise from the lymph nodes or spleen. The prognosis and classification of the various types is a very complex subject and this type of cancer doesn't always respond to chemotherapy or radiation therapy. It is more common in elderly people and those with poor immune systems and the main symptom is painless enlargement of lymph nodes.
3. ________________: Disorders of the immune system often referred to as 'atopic'. These reactions occur to normally harmless environmental substances and can only occur if the person has been previously exposed to the substance and developed antibodies to it. Symptoms range from mild (e.g. runny nose, watery eyes) to severe (e.g. anaphylactic shock).
4. _______: (Human Immunodeficiency Virus) ___________: (Acquired Immune Deficiency Syndrome): disorder of immune system (T4-cell destruction) whereby a virus attacks and leaves the body vulnerable to infection. It is transmitted by blood and body fluids.
5. ____________________________: (AKA infectious mononucleosis): An infectious disease caused by the 'Epstein Barr' virus affecting lymph nodes in the neck, armpits and groin. It mainly affects adolescents and is often called the 'kissing disease' as it can spread through saliva. Symptoms include swelling and tenderness of lymph nodes, fever, sore throat, lethargy and loss of appetite. It is diagnosed by the presence of a large number of monocytes in the blood.

6. ________________: This is an inflammation of the tonsils and may be viral or bacterial. It is characterised by sore throat, difficulty swallowing and fever.
7. ____: A chronic autoimmune connective tissue disease that can affect any part of the body. Symptoms vary but can include inflammation, painful joints, weight loss, hair loss, enlarged spleen, and a 'butterfly' rash across the cheeks. The cause is unknown but it affects young women mostly. A 'discoid' variety affects areas of the skin exposed to sunlight causing red scaly patches that leave pigmentation, atrophy and white scars. It can also affect the scalp, lips and inside the mouth.
8. _________________: Swelling caused by a buildup of lymph fluid when the lymph nodes do not drain properly, often in the legs. It can be due to a congenital defect in which there is a lack of lymph nodes or vessels or it can be caused by surgical removal of or damage to these. Parasites, tumours or injury to the lymph nodes or vessels can also cause it.
9. ________________: An abnormal production of leucocytes, which crowd out healthy cells in the bloodstream and invade other organs leading to increased susceptibility to infection, bleeding and anaemia.

Respiratory System Disorders

Fill the gaps with these words:

Tuberculosis (TB); Pulmonary embolism; Cystic fibrosis; Pneumonia; Asthma; Lung cancer; Rhinitis; Bronchitis; Sinusitis; Common cold; Cancer; Cor pulmonale; Tonsillitis; Emphysema; Cough; Pharyngitis; Influenza (flu); Laryngitis; Hay fever; Pleurisy; Chronic obstructive pulmonary disease/chronic obstructive airways disease (COPD/COAD); Hyperventilation; Whooping cough (pertussis); Pneumothorax; Pulmonary fibrosis; Sarcoidosis; Tracheitis; Severe acute respiratory syndrome (SARS); Snoring; Bronchiolitis

1. ________________: A chronic respiratory disease in which the airways unexpectedly and suddenly narrow, often in response to an allergen, cold air, exercise, or emotional stress. Symptoms include wheezing, shortness of breath, chest tightness, and coughing.
2. ________________: Inflammation of the bronchial tubes, causing coughing, shortness of breath and fatigue. It can result from infection, smoking, etc.
3. ________________: A malignant tumour caused by abnormal and uncontrolled cell division. It may spread to other parts of the body through the lymphatic system or the blood stream and can affect the lungs, throat or other parts of the respiratory system and is often associated with smoking.
4. __________________________: A very common, mild viral infection of the nose and throat with symptoms including sneezing, sniffling, a runny or blocked nose, a sore throat, coughing and headache. It is contagious and often spread by sneezing.
5. ______________: A sudden and often repetitively occurring reflex which helps to clear the large breathing passages from excess secretions, irritants, foreign particles and microbes.
6. ________________: Alveoli lose their elasticity, preventing effective breathing and gas exchange - results in coughing, shortness of breath and wheezing.
7. __________________: Allergic rhinitis due to an allergy to pollen. Symptoms include sneezing, runny nose and sore itching eyes.

8. ___________________: A highly contagious disease caused by various types of viruses and characterized by the sudden onset of fever, headache, exhaustion, muscle aches, cough, and sore throat.
9. __________________: An inflammation of the larynx. It causes hoarse voice or the complete loss of the voice because of irritation to the vocal cords in the larynx.
10. _________________: Inflammation of the pleura that may result in excess fluid. Causes localised pain, coughing and shortness of breath.
11. _________________: Inflammation of the throat or pharynx. In most cases it is painful, and thus is often referred to as a sore throat. It can be due to bacteria (streptococcus A) or virus.
12. _________________: Inflammation of the lungs due to infection, resulting in fever, pain, cough, fatigue, and shortness of breath - can be due to bacteria, virus or fungus. Those susceptible include smokers, the elderly, infants, immune compromised individuals and those with obstructive lung disease.
13. ______________________________: Blockage of a pulmonary artery or one of its branches, usually occurring when a deep vein thrombus (blood clot from a vein) becomes dislodged from its site of formation and travels to the arterial blood supply of one of the lungs.
14. _________________: Commonly known as a runny nose, this is the medical term describing irritation and inflammation of internal areas of the nose.
15. _________________: Inflammation of one of the paranasal sinuses.
16. _________________: Inflammation of the tonsils (usually palatine tonsils).
17. ____________________: Contagious disease of lung/s caused by airborne bacteria that destroys lung tissue, which is replaced by fibrous connective tissue and nodules called 'tubercles'. Results in fever, cough, night sweats, decreased energy and appetite leading to weight loss.
18. ___________________________: Inflammation of the membranes lining the bronchioles.
19. _____________________________: Enlargement of the right ventricle of the heart due to disease of the lungs or the pulmonary blood vessels or long term high blood pressure.
20. ____: Chronic bronchitis and emphysema, a pair of two commonly co-existing diseases of the lungs.

21. ______________________: An inherited disease in which thick mucus clogs the lungs and blocks the ducts of the pancreas.
22. ________________: Over-breathing to the extent that the blood carbon dioxide level is lowered and may lead to tingling in the extremities and dizziness.
23. ___________________: A disease of uncontrolled cell growth in lung tissue. This growth may lead to metastasis, which is the invasion of adjacent tissue and infiltration beyond the lungs. The most common symptoms are shortness of breath, coughing (including coughing up blood), and weight loss.
24. _______________________________: An infectious disease characterised by recurrent spasmodic bouts of coughing that continue until the breath is exhausted, ending with a characteristic 'whoop' as air is inhaled.
25. __________________: Collapse of one or both lungs caused by accumulation of gas or air in the pleural cavity, or resulting from injury or disease.
26. _________________________: Thickening and scarring of the lungs. This can occur after exposure to asbestos, silicates and other materials or most often without any known exposure or causes. This is a serious disorder that occurs mainly in older individuals and presents with a cough and shortness of breath with exertion, as the fibrotic tissue cannot carry out gas exchange efficiently.
27. ________________: A disease of unknown origin that causes small lumps (granulomas) to develop in a great range of body tissues due to chronic inflammation. It can appear in almost any body organ, but most often starts in the lungs or lymph nodes.
28. ___________________________________: A severe form of pneumonia, caused by a virus, that appeared in outbreaks in 2003.
29. _______________: The action or sound of breathing during sleep with harsh, snorting noises caused by vibrations of the soft palate.
30. ________________: A bacterial infection of the trachea that is capable of producing airway obstruction.

Digestive System Disorders

Fill the gaps with these words:

Cancer; Candida albicans; Cirrhosis; Colitis; Crohn's disease; Diverticulosis; Diverticulitis; Enteritis; Gastritis; Heartburn (reflux oesophagitis); Hernia; Hiccoughs; Indigestion (dyspepsia); Appendicitis; Anorexia (nervosa); Bulimia (nervosa); Irritable bowel syndrome (IBS); Constipation; Coeliac disease; Diarrhoea; Diabetes mellitus; Flatulence; Inflamed gallbladder; Jaundice; Stress; Ageusia (X2); Hypogeusia; Hypergeusia; Gingivitis; Gallstones; Ulcer; Ulcerative colitis; Pernicious anaemia; Oesophageal ulcer

1. ______________________________: Psychological condition characterised by loss of appetite, fear of gaining weight and refusing to eat. It predominately affects teenagers (particularly females) and can be fatal if not treated.
2. ____________________: Acute inflammation of the appendix, usually treated by removal.
3. ___________________________: Psychological condition – the opposite of anorexia. Presents as an insatiable hunger that includes binge eating followed by evacuation methods such as self-induced vomiting and excessive use of laxatives. It often affects teenagers and young women.
4. _____________: Any malignant growth or tumour caused by abnormal and uncontrolled cell division. It can occur anywhere in the digestive system.
5. _______________: Chronic disease of the liver caused by damage from toxins (including alcohol), metabolic problems, hepatitis or nutritional deficiencies. It is characterised by an increase of fibrous tissue and the destruction of liver cells.
6. ________________: Excessive bile pigment in the blood causing the skin to look yellow. It is caused by a malfunctioning gall bladder or obstructed flow of bile.
7. ______________: Can cause an overproduction of gastric juices and is a common cause of ulcers.

8. The complete inability to taste is called ________________; the reduced ability to taste is called ________________ and the enhanced ability to taste is called ________________. ________________ is rare because there are three different nerves that carry taste information to the brain.
9. ____________: Erosion of the wall of the stomach or intestines often caused by high acid levels. It can be exaggerated by stress, which causes excess stomach acid. Those occurring in the mouth (apthous) have various causes including iron or vitamin deficiency, stress or injury e.g. biting the inside of the mouth.
10. ________________: Bile pigments combined with cholesterol and calcium salts forming stones in the gall bladder.
11. ______________: Inflammation of the gums often caused by a buildup of food particles and plaque (a sticky film containing bacteria) on and around the teeth.
12. ____________________________: Back-flow of acidic stomach contents into the oesophagus causing a burning sensation.
13. __________: Part of the body protruding abnormally through a tear or opening in an adjacent part; e.g. inguinal groin - when there is a weak spot in the area where the abdomen meets the thigh; hiatus - resulting from protrusion of part of the stomach through the diaphragm.
14. ________________: A spasmodic contraction of the diaphragm that repeats several times per minute. In humans, the abrupt rush of air into the lungs causes the epiglottis to close, creating the 'hic' sound.
15. ________________: A disorder of digestive function characterised by discomfort, heartburn or nausea.
16. ____________________________________: A disorder of unknown origin (although stress and poor diet can contribute) causing stomach and bowel pain with constipation and diarrhoea.
17. ________________: Infrequent or uncomfortable bowel movements, causing hard faeces to block the rectum. It can be due to lack of fibre, lack of fluids, lack of exercise and is sometimes caused by stress.
18. ____________________________: An autoimmune disease characterised by sensitivity of the lining of the small intestine to gluten, causing destruction of the villi and failure to digest food properly.

19. ________________: Frequent bowel evacuation or the passage of abnormally soft or liquid faeces. It can be due to intestinal inflammation, anxiety or irritable bowel syndrome. If severe or prolonged it can lead to the loss of body fluids, salts and nutrients in the faeces.

20. ______________________________: A metabolic disease caused by failure of the pancreas to produce insulin, a hormone that allows blood sugar (glucose) to be taken up by cells that require it for function.

21. ________________: The release of a mixture of gases known as flatus produced by symbiotic bacteria and yeasts living in the gastrointestinal tract. It is released under pressure through the anus, often accompanied by odour and sound.

22. _______________: A fungus that causes yeast infections like thrush in the mouth and throat as well as the intestines and other parts of the body. It is actually part of the normal flora of the mouth, intestines, and vagina, but can overgrow and cause disease (candidiasis, or thrush), especially in those with a weak immune system.

23. ______________: Inflammation of the large intestine that has clinical signs including diarrhoea and abdominal pain. There are many forms including ulcerative, infectious, pseudomembranous, and spastic.

24. _______________________________: A serious disease that causes ulcers and irritation in the inner lining of the colon and rectum.

25. ________________________________: A chronic form of inflammatory bowel disease that usually affects the lower small intestine (called the ileum) or the colon, but it can affect the entire gastrointestinal tract. It is thought to be an autoimmune condition.

26. ___________________: A condition that occurs when small pouches (diverticula) push outward through weak spots in the colon.

27. ____________________: A condition that occurs when small pouches in the colon (diverticula) become infected or irritated.

28. ________________: Inflammation of the intestines (especially the small intestine) due to viral infection and usually characterised by diarrhoea.

29. ________________: Inflammation of the lining of the stomach causing nausea, loss of appetite and discomfort after eating.

30. ________________________: Inflammation of the gallbladder usually caused by gallstones. It may settle down with treatment or may require removal of the gallbladder.

31. ______________________________: A chronic progressive anaemia of older adults, thought to result from a lack of intrinsic factor (a substance secreted by the stomach that is responsible for the absorption of vitamin B12).

32. ____________________________________: A sore or erosion of the oesophagus generally caused by excessive exposure to acid.

Nervous System Disorders

Fill the gaps with these words:

Alzheimer's disease; Concussion; Depression (clinical depression, bipolar affective disorder, seasonal affective disorder (SAD), post-natal depression); Headache; Meningitis; Migraine; Myalgic encephalomyelitis (ME); Motor neurone disease; Myasthenia gravis; Neuralgia; Neuritis; Paralysis; Parkinson's disease; Peripheral neuropathy; Poliomyelitis; Sciatica; Stroke; Transient iaschemic attack (TIA); Bell's palsy; Drug abuse; Epilepsy; Alcohol abuse; Cerebral palsy; Multiple sclerosis; Spina bifida

1. ______________________________: Depresses the central nervous system and long term effects include damage to the frontal lobes of the brain, an overall reduction in brain size and various health problems.
2. ________________________: Facial paralysis caused by injury or infection affecting the facial nerve (cranial nerve V11).
3. __________________________: Damage to the motor area of the brain usually during or before birth, causing motor system control problems, abnormal posture, muscle spasticity and speech impairment. It is not a progressive condition.
4. __________________: A medical condition in which the sufferer experiences seizures (convulsions) and blackouts. Various smells or flashing lights can trigger it.
5. ________________________: In high doses or taken over long periods of time, this can cause the brain's neurological communication to become permanently rewired, or the brain can stop producing essential chemicals needed for normal functioning. It can also cause irregular heartbeat, high blood pressure and respiratory problems.
6. ________________: This is a symptom of a number of different conditions of the head and sometimes affecting the neck. Some causes are benign while others are medical emergencies. It is among the most common pain complaints.
7. __________________: A severe, disabling headache, usually affecting only one side of the head, and often accompanied by nausea, vomiting, photophobia and visual disturbances. The causes are not well understood but triggers include food, hormonal changes, sensations (e.g. light) and stress.

8. ______________________________: A chronic slowly progressive disease of the motor nerves marked by wasting of muscles and associated with weakness and paralysis. It does not affect the senses.
9. ______________________________: A disease that is due to loss of the myelin sheath in the CNS. It causes muscular weakness, loss of co-ordination and sensation problems.
10. ____________________: A mental state characterised by a pessimistic sense of inadequacy and despondent lack of activity. It includes the following (11-14 below):
11. ______________________________: Serious enough to require psychiatric intervention and treatment.
12. ______________________________: Previously known as manic depression this condition affects moods, which can swing from one extreme to another.
13. ______________________________: Caused by the lack of natural light during the winter months.
14. ______________________________: A more severe form than 'baby blues,' which can make many mothers feel miserable, weepy, tired, on edge or worry about the baby or themselves and can be caused by changes in hormone levels. It usually disappears within a day or so, but some women experience a more severe form with symptoms that persist in varying degrees for a number of years.
15. ______________________________: Defined as 'inflammation of the brain and spinal cord with muscle pain' and also known as chronic fatigue syndrome or post viral fatigue. This is a poorly understood, variably debilitating disorder of uncertain cause. Symptoms include general exhaustion as well as aches and pains in muscles.
16. ________________: Inflammation of a nerve due to injury or infection, causing pain along the length of the nerve or a loss of control of structures affected by that nerve.
17. ________________: General term for pain along the course of a nerve.
18. ______________________________: A progressive disease of the basal ganglia in the brain, resulting in dopamine (neurotransmitter) loss. It affects the motor functions of the cerebellum and causes rigidity in muscles, tremors and gait disturbances.
19. ________________: Neuralgia of the sciatic nerve (gluteal region).

20. ______________: A sudden loss of consciousness due to lack of oxygen to the brain caused by the rupture or occlusion of a blood vessel to the brain.

21. ____________________________________: Sometimes called a mini-stroke, this is a temporary restriction of blood supply to the brain causing short-term symptoms such as vision loss or impairment. It often occurs for only a few minutes and is usually resolved within 24 hours. It is a sign that a full stroke may occur so it should be investigated.

22. ______________________________: A progressive degenerative disease of brain tissue causing impaired memory, thinking and behaviour. It is one of the most common causes of dementia. There is significant loss of intellectual ability and memory capacity that is often severe enough to interfere with social or occupational functioning, and there can be episodes of paranoia and mood change. The cause is unknown but involves loss of some neurones and development of plaques on some neurones along with the entanglement of protein filaments within neurones.

23. ________________: Injury to the brain caused by a blow to the head, usually resulting in headaches, nausea and sometimes loss of consciousness.

24. ________________: Infectious disease characterised by inflammation of the meninges (membrane that surrounds the brain and spinal cord) caused by a bacterial or viral infection. Symptoms include headache, stiff neck, fever and nausea.

25. ____________________________: A disorder affecting the space between a nerve and a muscle (neuromuscular junction) that results in transient motor weakness of the face and limbs. It is due to an autoimmune process affecting the chemical acetylcholine.

26. ________________: Complete loss of muscle function for one or more muscle groups. It can cause loss of feeling or mobility in the affected area.

27. ________________________________: Numbness and/or tingling in the fingers or toes. It can occur as a side effect of some chemotherapy drugs but can have other causes such as diabetes or lack of vitamin B12.

28. ________________: An acute viral infectious disease spread from person to person that affects motor neurones in the spinal cord and brainstem, leading to muscle weakness, paralysis and sometimes deformity. It is rare nowadays due to vaccination.

29. ________________________: (Latin: 'split spine). This is a developmental birth defect resulting in a cleft in the spinal column through which the spinal membranes and spinal cord may protrude. It is characterised by hydrocephalus - an abnormal condition in which cerebrospinal fluid collects in the ventricles of the brain. In infants it can cause abnormally rapid growth of the head, bulging fontanel, a small face and other neurological disorders.

Endocrine System Disorders

Fill the gaps with these words:

Cushing's syndrome; Cushing's disease; Diabetes insipidus; Diabetes mellitus; Gigantism; Goitre; Hypothyroidism; Hyperparathyroidism; Hypoparathyroidism; Insomnia; Myxoedema; Acromegaly; Addison's disease; Grave's disease; Hyperthyroidism (thyrotoxicosis); Polycystic ovarian syndrome (PCOS); Stress

1. ___________________________: This is caused by hyposecretion of adrenocortical hormones (particularly cortisol but sometimes also aldosterone). The effects are weakness, muscular atrophy, gastric problems, low blood pressure, dehydration and dark pigmentation of the skin.
2. ____________________________: Due to hypersecretion of adrenocortical hormones. The effects are weight gain, reddening of face and neck, excess body and facial hair, raised blood pressure, loss of mineral from bones (osteoporosis), 'moon face', deposition of fat on upper back (buffalo hump), muscle wasting in limbs, raised blood glucose and sometimes mental disturbance. __________________________________ is the term used when a tumour in the pituitary gland causes the syndrome.
3. ______________________________: This is a condition characterised by frequent and heavy urination, excessive thirst and an overall feeling of weakness. This condition results from a deficiency of vasopressin (ADH), the pituitary hormone that regulates water balance in the kidneys. In this condition, blood glucose levels are normal.
4. ____________________________: This is a metabolic disease caused by failure of the pancreas to produce insulin, the hormone that allows blood sugar (glucose) to be taken up by cells that require it for normal function.
5. ______________: An abnormally enlarged thyroid gland that can result from underproduction or overproduction of thyroid hormones or from a deficiency of iodine in the diet.

6. ______________________: This is an autoimmune disease that affects the thyroid, causing it to enlarge to twice its size or more, become overactive, with related hyperthyroid symptoms such as increased heartbeat, muscle weakness, disturbed sleep, and irritability. It can also affect the eyes, causing bulging eyes (exophthalmos). It affects other systems of the body, including the skin, heart, circulation and nervous system.
7. __________________: This is a clinical condition in which there is pathologically excessive production of thyroid hormones. The most common form is Grave's disease.
8. __________________: An underactive thyroid with symptoms including weight gain, constipation, dry skin, tiredness, slow metabolism and sensitivity to the cold.
9. ________________: This is a symptom of any of several sleep disorders, characterised by persistent difficulty falling asleep or staying asleep despite the opportunity. It is a symptom, not a stand-alone diagnosis or a disease, and can be due to stress, low melatonin levels (from pineal gland) or other factors.
10. ____: Also called Stein-Leventhal Syndrome, this is a hormonal disorder characterised by incomplete development of Graafian follicles sometimes due to inadequate secretion of luteinising hormone but can have other causes. The follicles fail to ovulate and remain as multiple cysts distending the ovary.
11. ________________: This is severely advanced hypothyroidism marked by dry skin and swelling around the lips and nose, thickening of skin on the body e.g. legs, and mental deterioration.
12. ___________: This causes the body to respond as to any physical danger ('fight or flight' response). It causes the adrenal medulla to release adrenaline and noradrenaline. The result of this is the same as the sympathetic autonomic nervous system, i.e. increased heart rate and breathing, glucose released from the liver, blood supply increases to skeletal muscles, etc.
13. ______________: A chronic disease marked by enlargement of the bones and connective tissue of the extremities, face, and jaw that is caused by over-activity or tumour of the pituitary gland.

14. ___________________: A condition where there is overproduction of growth hormone by the pituitary gland in a child before the bone growth plates close, resulting in excessive long bone growth, accompanied by muscular weakness and often sexual dysfunction.

15. _________________________: This is excessive secretion of parathyroid hormone resulting in abnormally high levels of calcium in the blood. It can affect many systems of the body e.g. it may cause or worsen many problems including kidney stones, depression, pancreatitis, fatigue, constipation, osteoporosis or joint pains.

16. __________________________: This is inadequate secretion of parathyroid hormone resulting in abnormally low levels of calcium in the blood. Signs and symptoms include tingling lips, fingers, and toes, muscle cramps, pain in the face, legs, and feet, muscle spasms called tetany (can lead to spasms of the larynx, causing breathing difficulties), convulsions (seizures).

Urinary System Disorders

Fill the gaps with these words:

Urinary retention; Glomerulonephritis; Prostatitis; Benign prostatic hyperplasia (BPH); Pyelonephritis; Renal failure; Gout; Urinary incontinence; Kidney stones; Proteinuria; Interstitial cystitis (IC); Diabetes insipidus; Cancer; Urinary tract infections (UTI's); Cystitis; Nephritis (Bright's disease)

1. ____: A condition affecting the prostate gland in males. The prostate enlarges and can interfere with urinary function (particularly in older men) by squeezing the urethra, making it difficult to urinate. Men with this condition often have other bladder symptoms including an increase in frequency of urination both day and night. Most men over the age of 60 have this problem to some degree, but not all have problems with blockage.
2. ___________: Signs include blood in urine, pain on urination, pelvic/abdominal pain and genital swelling - any of these symptoms should be medically investigated.
3. ____________: Inflammation of the bladder, causing pain on urination, often caused by infections, more common in women due to a shorter urethra.
4. ____________________________: A rare disease resulting from a deficiency of ADH also known as vasopressin (the pituitary hormone that regulates fluid balance in the kidneys). It is characterized by the chronic excretion of large amounts of pale dilute urine, resulting in dehydration and extreme thirst.
5. _________: Symptoms include painful swelling and inflammation in one or more of the joints. It usually affects the big toe, but can develop in any joint of the body. This condition is caused by a buildup of uric acid – a waste product that is produced during the process of metabolism. When the kidneys cannot excrete uric acid properly, high levels accumulate in the blood and uric acid crystals form in the joints, causing pain and inflammation.
6. __________________________: Loss of bladder control and the involuntary passage of urine. There are many causes and types and many treatment options. Treatments range from simple exercises to surgery. Women are affected more than men.

7. ___________________________________: A chronic bladder disorder also known as painful bladder syndrome and frequency-urgency-dysuria syndrome. In this disorder, the bladder wall can become inflamed and irritated. The inflammation can lead to scarring and stiffening of the bladder, decreased bladder capacity, spot bleeding, and, in rare cases, ulcers in the bladder lining. The cause is unknown.
8. ______________________________: Deposits of substances (urea, calcium, etc.) forming solid stones within the renal pelvis, bladder or ureters and sometimes needing surgical removal.
9. _________________: Inflammation of the kidneys and often referring to a range of inflammatory disorders of the kidneys characterised by impaired kidney function, fluid and urea retention and blood in the urine (uraemia). It includes the two disorders below in 10 & 11.
10. ___________________________: Bacterial infection of the kidneys characterised by fever, shivering and pain.
11. __________________________________: Nephritis marked by inflammation of the glomeruli (blood capillaries) of the kidneys. It is characterised by oedema, decreased production of urine and the presence of blood and protein in urine.
12. ________________________: Inflammation of the prostate gland resulting in urinary frequency and urgency, burning or painful urination (dysuria), and pain in the lower back and genital area. In some cases, it is caused by bacterial infection and can be treated with antibiotics.
13. __________________________: The presence of abnormal amounts of protein in the urine. Healthy kidneys take waste out of the blood but leave in protein. Protein in the urine does not cause a problem by itself, but may be a sign that the kidneys are not functioning properly. Pregnant women are routinely screened for this condition because its development is associated with preeclampsia (also known as toxaemia of pregnancy). Preeclampsia is a pregnancy-specific disorder where this condition and hypertension develop at the same time. Symptoms can include oedema (swelling), nausea, and headaches during pregnancy. In rare cases, it can cause severe symptoms such as seizures. Preeclampsia can be dangerous for both the mother and baby.

14. ______________________________: This is a bladder-emptying problem, a common urological problem with many possible causes. Normally, urination can be initiated voluntarily and the bladder empties completely but this condition is the abnormal holding of urine. Treatment will depend on the exact cause.

15. ______________________: This results when the kidneys are not able to regulate water and chemicals in the body or remove waste products from the blood. Acute conditions can be caused by an accident that injures the kidneys, loss of a lot of blood, or some drugs or poisons, and may lead to permanent loss of kidney function. If the kidneys are not seriously damaged, they may recover. A chronic condition is the gradual reduction of kidney function that may lead to permanent kidney failure, or end-stage renal disease (ESRD). A person may go several years without knowing they have this condition.

16. ____________________________________: Usually caused by bacteria in the urinary tract. Women get these types of infections more often than men and they are usually treated with antibiotics. Drinking lots of fluids also helps by flushing out the bacteria. The exact name of the infection depends on its location in the urinary tract.

Reproductive System Disorders

Fill the gaps with these words:

Chlamydia; Cysts; Ectopic pregnancy; Endometriosis; Fibroids; Galactorrhea; Hysterectomy; Mastitis; Mastalgia; Menstrual disorders; Polycystic ovarian syndrome (PCOS); Pre-menstrual syndrome; Prostatitis; Vulvovaginal candidiasis (thrush)

1. ____________: Most women develop breast pain at some stage in life. In most cases the pain develops in the days just before a period and it is thought to be due to having breast tissue that is more sensitive than usual to the normal hormone changes that occur each month. In some cases the pain is not related to periods and can be due to muscular or bony problems of the chest wall. Pain of unknown origin should be investigated.
2. ____________: A benign one usually develops in the breast due to dilated ducts that create fluid-filled sacs in the breasts. Fibroadenomas are breast masses resulting from excess growth of the glandular and connective tissue and are most common among younger women or those who are pregnant. A health care professional should be consulted to determine the source of any breast lumps and rule out the possibility of a malignant tumour.
3. ________________: This is the spontaneous flow of milk from the breasts, unassociated with childbirth or nursing. It can be due to deregulation of certain hormones such as prolactin or local causes such as excessive nipple stimulation.
4. ______________: Inflammation, swelling and tenderness in the breast(s) of a breastfeeding woman that may be caused by a blocked milk duct or an infection.
5. ________________: This is a common sexually transmitted disease (STD) caused by the trachomatis bacterium that can damage a woman's reproductive organs. Even though symptoms are usually mild or absent, serious complications that cause irreversible damage, including infertility, can occur 'silently' before a woman ever recognises a problem. The disease also can cause discharge from the penis of an infected man.

6. ______________________________: This is a pregnancy that begins to develop outside the uterus i.e. the fertilised ovum develops in a fallopian tube leading to danger of haemorrhage or death if not treated.
7. ____________________: This is a condition where tissue that normally lines the uterus grows outside the uterus in other areas of the body. It can cause pain, irregular menstrual bleeding, and infertility for some women.
8. ________________: These are benign growths or tumours within the uterus and are the most common tumours found in women. Many women often have no symptoms, but some can develop heavy bleeding, long periods, pelvic pressure or pain, urinary problems, or constipation.
9. _________________________________: These include **amenorrhea** - absence of menstruation which has various causes; **dysmenorrhea** - caused by spasm or congestion of the uterus, imbalance of hormones or stress, resulting in difficult or painful menstruation, **menorrhagia** - abnormally heavy menstrual cycle with various causes.
10. ___________________________________: Physical and psychological symptoms occurring the week before a woman's menstrual period. Symptoms may include bloating, headache, irritability, anxiety, depression, low self-esteem, difficulty sleeping, changes in appetite, fatigue, breast swelling and tenderness. It has no exact cause but theories include abnormal sensitivity to hormones released prior to menstruation, magnesium/calcium imbalance and lack of vitamin B6.

11. ____: This is a hormonal disorder characterised by incomplete development of Graafian follicles. The exact cause is not clear but in some cases it is due to inadequate secretion of luteinising hormone (LH), although a high level of LH is found in about 4 in 10 sufferers. Some women suffer but do not develop the full-blown syndrome. Women with this condition often have 'insulin resistance' i.e. cells in the body are resistant to the effects of a normal level of insulin and more insulin is produced to keep the blood sugar normal. This then causes the ovaries to make too much testosterone. A high level of insulin and testosterone interferes with normal development of follicles in the ovaries. As a result, many follicles tend to develop but often do not develop fully. This causes problems with ovulation and hence period problems and reduced fertility. Increased testosterone levels in the blood cause excess hair growth on the body and thinning of scalp hair. Increased insulin also contributes towards weight gain.

12. __________________: This is inflammation of the male prostate gland, with associated pain and discomfort. It is categorised by being acute (sudden onset) or chronic (ongoing), bacterial or non-bacterial.

13. _____________________________________: This is a yeast infection of the lower female reproductive tract causing vaginal discharge, irritation, and itching. Sometimes male partners can suffer candidal balanitis - inflammation of the end of the penis. Causes include taking antibiotics or steroid medications, HIV infection or AIDS, receiving chemotherapy for cancer or drugs to suppress the immune system. Avoidance of sugar and yeast in the diet can help.

14. ____________________: This is the surgical removal of the uterus and sometime includes removal of the ovaries (oophorectomy).

Anatomical Terminology

Across

3. Pectoral relates to which part of the body (5)
6. The study of disease is called (9)
8. Something that increases a person's susceptibility to disease (4, 6)
11. A disease spread from animal to humans (8)
12. Signs and symptoms indicate a more serious pathology (3,4)
13. This part of the body is in the appendicular skeleton (6)
14. Signs and symptoms that characterise a particular medical condition (8)

Down

1. On opposite sides of the body (13)
2. A disease of unknown origin is termed (10)
4. In the centre, above the navel and over the stomach (10)
5. A plane diving the body into right and left regions (8)
7. A plane dividing the body into superior & inferior regions (10)
9. A disease of long duration (6)
10. The pH (acidity/ alkalinity) of water is (5)

Body Compounds

Across

2. Vitamin A is also called (7)
4. Type of anaemia that affects people who lack Vitamin B12 (10)
10. Sugars and starches used as fuel by the body (13)
11. Necessary for thyroid hormone production (6)
13. Water has a pH of (5)
14. Needed for healthy bones and teeth (7)

Down

1. Excess can cause issues with blood pressure (6)
3. Vitamin D is derived from this (8)
5. A critical ingredient in haemoglobin (4)
6. A component of water, all foods and most organic molecules (8)
7. Lipids are also known as (4)
8. Needed for cellular respiration (6)
9. Makes up about 60% of human body weight (5)
12. Compounds containing carbon atoms (7)

Histology

Across

6. Stable chromosome ends that protect DNA (9)
8. Study of the structure and function of cells and tissues (9)
9. When chromosomes do not divide properly (14)
11. Movement of water across a cell membrane (7)
12. Final stage in cell division (9)
13. 'Cell eating' to absorb solid materials such as bacteria (12)
14. Cell division stage when chromosomes line up in the centre (9)

Down

1. Hold all hereditary information (5)
2. Organelles that produce energy inside cells (12)
3. Waste disposal organelles (9)
4. Organelles involved in cell division (10)
5. An unusual arrangement of chromosomes (13)
7. System of fibrous connective tissue throughout the body (6)
10. Organelles that make proteins (9)

Skin

Across	Down
2. Sweat glands in the underarms (8)	**1.** Cells that clear 'debris' when skin is injured (11)
4. Excessive sweat production (13)	**3.** Vitamin D3 is also called (15)
7. The skin's natural oil (5)	**5.** Skin that lacks water rather than oil (10)
8. Collagen & elastin are found here (6)	**6.** Cells found in subcutaneous layer of skin (7)
10. Epidermal layer where cells divide (5)	**9.** Protects the skin from the sun (7)
12. Keeps the skin waterproof (7)	**11.** Sweat glands found all over the body (7)
13. Foul body odour from sweat glands (12)	
14. Blood vessels in the dermis do this when body is cold or stressed (9)	

Skin Disorders

Across

3. Malformed, dilated blood vessels (11)
6. Most serious form of skin cancer (8)
10. Inflammatory autoimmune disease (5)
12. Zoster virus that causes chicken pox (9)
13. Also known as herpes zoster (8)
14. Results from increased epidermal cell turnover (9)

Down

1. An autoimmune disease in which the skin thickens (11)
2. Bacteria, resistant to certain antibiotics (4)
4. Caused by the human papillomavirus (5)
5. Fatty deposits under the skin or around tendons (9)
7. Causes the skin to become inflamed or irritated (6)
8. Highly contagious skin infection - mainly affects children (8)
9. Loss of melanocytes in areas of the skin (8)
11. Brown pigmentation marks, often on the face (7)

Hair

Across	Down
2. Soft, downy hair of foetus (6)	**1.** Term for hair loss (8)
7. Central part of a hair (7)	**3.** Active growth phase of hair (6)
9. Pigment that gives hair its colour (7)	**4.** Also known as pediculosis capitis (8)
10. Protein that makes hair waterproof (7)	**5.** Area of active cell division in hair bulb (6)
12. Middle layer of the 3 layers of a hair (6)	**6.** Resting phase of hair growth (7)
13. Transitional hair growth phase (7)	**8.** Outer part of a hair (7)
14. Excessive facial hair growth in women (9)	**11.** Part of hair above surface of skin (5)

Nails

Across

2. Spoon shaped, concave nails (11)
6. Technical term for nail disease (9)
7. Skin overlapping sides of nail plate (11)
9. Blue colour in fingernails due to lack of oxygen (8)
11. Dry, brittle, splitting nails (13)
12. Commonly known as the cuticle (10)
13. Causes pinprick holes in nails (9)
14. Medical term for nail biting (12)

Down

1. Fungus causing tinea ungium (8)
3. Nail infection caused by herpes simplex virus (7)
4. Seal underneath 'free' edge of nail (11)
5. An in-growing nail (15)
8. 'Half-moon' of nail root (6)
10. Protein that makes up nail plate (7)

Skeletal System

Across	Down
2. Fracture where bone is broken in several places (10)	**1.** The study of joints (10)
4. Joint inflammation (9)	**3.** Softening of the bones (12)
7. Freely movable joints (8)	**5.** Cells that break down bone (11)
9. Canals in bone carrying nerves & blood vessels (9)	**6.** Medical name for lower jaw (8)
12. Uric acid crystals in joints (4)	**8.** Process of bone formation (12)
13. Hard, dense bone tissue (7)	**10.** Bone tissue that makes bone marrow (6)
14. Outer skin of bone (10)	**11.** Condition where bones lose density (12)
15. Common name for zygomatic bones (10)	

Muscular System

Across

3. Widespread muscle and joint pain and fatigue (12)
5. Movement away from the body (9)
7. Neurotransmitter involved in muscle movement (13)
8. Disorders of nerves and muscles (13)
10. Temporarily paralyses muscles (5)
11. Thickest part of a muscle (5)
12. One of 2 proteins found in muscles (6)
13. Muscle contractions involving no movement (9)
14. Muscle tissue of internal organs (6)

Down

1. An individual muscle cell (9)
2. Muscle found in the heart (7)
4. The ability to be stretched without tearing (13)
6. Inflammation of muscle (8)
9. Connective tissue covering muscles (9)

Skeletal Muscles

Across	Down
2. Strongest tendon in the body (8)	**1.** Muscles that draw scapulae together (9)
9. Main muscle of mastication (8)	**3.** Muscles involved in smiling (9)
11. One of the calf muscles (6)	**4.** Tiny muscle at bridge of the nose (8)
12. Muscles that flex the knee (10)	**5.** Broad, flat muscle at front of neck (8)
13. Shoulder muscle that abducts arm (7)	**6.** Muscle involved in sciatica (10)
14. Muscles that extend the knee (10)	**7.** Strong hip flexor muscles (9)
	8. Muscles affected by tennis elbow (9)
	10. Large muscle of upper back (9)

Cardiovascular System

Across

5. Cells that eat pathogens (10)
10. T-lymphocytes mature here (6)
11. White blood cells that kill parasites (11)
12. Person who discovered blood types (11)
13. Pressure felt when heart muscle contracts (8)
14. Caused by low haemoglobin levels (7)

Down

1. Arteries in neck where pulse is felt (7)
2. Carries oxygen on red blood cells (11)
3. Kidney hormone that stimulates red blood cell production (14)
4. Also known as the sino-atrial node (9)
6. Narrowing of arteries due to fatty deposits (15)
7. Cells involved in blood clotting (9)
8. Valve between right atrium and ventricle (9)
9. Caused by extreme response to infection in bloodstream (6)

Lymphatic System & Immunity

Across

3. Fever caused by the 'Epstein Barr' virus (9)

7. Underarm lymph nodes (8)

10. This part of the immune system remembers pathogens (8)

11. Causing immune system to produce antibodies (8)

12. This organ filters blood (6)

13. Lymph nodes in elbow creases (14)

14. Cancers of the lymphatic system (9)

Down

1. Part of immune system present at birth (6)

2. Produced by B-cells (10)

4. Lymphatic capillaries in the small intestine (8)

5. This virus weakens the immune system (3)

6. Swelling caused by buildup of lymph in tissues (10)

8. Lymph nodes behind knees (9)

9. Life-threatening reaction to allergens (11)

Respiratory System

Across

2. Also known as voice box (6)
6. Inflammation of inner lining of nose (8)
9. Spaces within bone (7)
10. Nerve controlling diaphragm (7)
12. Respiratory centre of brain is here (7)
13. Membrane surrounding lungs (6)
14. Cartilage flap that prevents food entering trachea (10)

Down

1. Also known as whooping cough (9)
3. Produced in allergic reactions (9)
4. Infection that inflames the alveoli (9)
5. Disease once called consumption (12)
7. COVID-19 is caused by this type of virus (11)
8. Condition in which airways narrow and swell often in response to an allergen (6)
11. Gas exchange occurs here (7)

Digestive System

Across

3. First section of small intestine (8)

5. Causes bulging pouches in the large intestine (14)

8. Nerve connection brain and gut (5)

10. Enzyme that breaks down fats (6)

11. Sphincter between oesophagus and stomach (7)

13. Inflammation of the small intestine (9)

14. Soft ball of chewed food (5)

Down

1. Large membrane lining the abdominal cavity (10)

2. Teeth that cut food and push it to the back teeth (8)

4. Viral infection causing liver inflammation (9)

6. Acid produced in stomach (12)

7. Taking solid and liquid food into the body (9)

9. Sphincter between stomach and small intestine (7)

12. Enzyme that breaks down carbohydrates (7)

Nervous system

Across

2. Nerve for sense of smell (9)
5. Also called the involuntary nervous system (9)
8. Part of nervous system involved in stress response (11)
10. Neurotransmitters that inhibit pain signals (10)
12. Also called the 'little brain' (10)
13. Also called the voluntary nervous system (7)
14. Membrane around brain & spinal cord (8)

Down

1. Neurotransmitter lacking in Parkinson's sufferers (8)
3. Nerve chemicals (17)
4. Nerve involved in mastication (10)
6. Fatty substance that insulates nerve axons (6)
7. Part of brain that stores long-term memories (11)
9. Group of muscles innervated by a particular spinal nerve (7)
11. Sharp pain that follows the path of a nerve (9)

The Senses

Across

2. The ear is located in this bone (8)
8. A savoury or meaty taste (5)
9. Receptors that detect sensations related to temperature (15)
11. Cranial nerve of hearing & balance (17)
12. Collective name for bones of middle ear (8)
13. Part of retina responsible for detailed central vision (6)
14. Causes lens of eye to become opaque and blur vision (8)

Down

1. Condition of increased pressure within the eyeball (8)
3. Another name for pain receptors (11)
4. The somatosensory system is involved with this sense (5)
5. Clear, thin layer covering front of eyes (6)
6. Tube between middle ear and pharynx (9)
7. Loss of sense of smell (7)
10. How many types of taste buds on the tongue? (5)

Endocrine System

Across

2. Anti-inflammatory hormone from adrenals (8)
5. Stimulates the production of breast milk (9)
10. Hormone involved in 'fight or flight' (10)
11. Diabetes caused by insulin problems (8)
12. The Islets of Langerhans is found here (8)
13. Triggers contractions in childbirth (8)
14. Helps maintain body's salt and water levels (11)

Down

1. The smooth running of body functions (11)
3. Produces hormones that regulate metabolic rate (7)
4. Also called anti-diuretic hormone (11)
6. Diabetes caused by insufficient ADH (9)
7. Enlargement of the thyroid gland (6)
8. Stimulates release of calcium from bones to blood (12)
9. Secretes the hormone melatonin (6)

Urinary System

Across

2. Substance found in urine (4)
4. Method of blood filtration using a machine (8)
9. Tubes leading from kidneys to bladder (7)
11. Also called Wilms' tumor, this rare kidney cancer mainly affects children (14)
12. Dilated part of the ureter inside kidney where urine collects is the renal? 6)
13. Common urinary tract infection in females (8)
14. Loss of bladder control (12)

Down

1. Filtering units in kidneys (8)
3. Another word for urination (11)
5. Cluster of capillaries that filters blood in the kidneys (10)
6. Inflammation of prostate gland (11)
7. Commonly known as bed-wetting (8)
8. Kidney hormone that stimulates red blood cells (14)
10. Pregnant women are routinely screened for this (11)

Reproductive System

Across	Down
3. Another name for reproductive cells (7)	**1.** Membrane covering external vaginal opening (5)
8. Abnormal, benign growths in uterus (8)	**2.** Inflammation of breast tissue (8)
10. High blood pressure & protein in urine during pregnancy (12)	**4.** Surgical removal of the uterus (12)
11. Lining of uterus (11)	**5.** Female canal leading from vulva to uterus (6)
13. Cell formed from union of ovum and sperm (6)	**6.** Tail of sperm that helps it move (9)
14. Produces progesterone and oestrogen during pregnancy (8)	**7.** Area of testes where sperm develop (10)
15. Absence of menstrual periods (10)	**9.** Pregnancy the develops in fallopian tubes (7)
16. Neck of the uterus (6)	**12.** Severe menstrual pain (12)

Study Tips for Students

Schedule study time: At the start of your course, build time for studying into your weekly routine so that you are not stressed with cramming it all in just before an exam.

Mnemonics: Studying anatomy and physiology involves a lot of remembering, and many students find that the use of mnemonics can make this easier. For example, taking the first one or two letters of each word from a list to create an acronym e.g. **S**imon **L**eft **T**he **P**arty **T**o **T**ake **C**athy **H**ome for remembering the names of the carpal bones of the wrist (**S**caphoid, **L**unate, **T**riquetral, **P**isiform, **T**rapezium, **T**rapezoid, **C**apitate, **H**amate). Another tip is to use the rhythm of an already familiar song, add the words you need to remember and sing it until it sticks!

Repetition: The best way to memorise all the different anatomical structures and their functions is by repetition. The more you go over something, the more likely you will be to remember it. Many students find re-writing notes to be helpful.

Note taking: Taking clear notes can help with remembering the material - whether taking notes in class or while reading a text - combining seeing the material with writing it down can increase your ability to retain the information. As well as underlining and highlighting important material in textbooks and study guides, it can be helpful to write down what you need to remember in your own words and re-read it until it makes sense.

Variety: Find a variety of ways to revise things e.g. write notes, use flashcards and colouring books, study visual diagrams and try to draw the anatomical structures and annotate them with facts and features.

Study buddy: Having a person to study with and test each other's knowledge can be very helpful for some people You can also get together with a 'study group' of friends and have a quiz based on the tests in this book! Having someone else to bounce ideas off of, or someone to quiz you can help with retaining information and identifying areas of weakness. Having a study group can also motivate you to study and keep you accountable.

Visual aids: An anatomically correct skeleton model will enable you to clearly see the bones and joints of the human body and will add a new dimension to your study. A human skeleton model with moving parts illustrates range of motion, while removable parts allow for more in-depth examination. Some educational models go beyond just bones and include nerves, arteries and intervertebral discs, as well as labelled areas for insertion points of muscles. Skeleton models come in different sizes e.g. full size, compact and desktop size, and can be a useful tool in your professional practice.

Anatomy colouring books and flashcards: Using colouring books is a great way to learn and memorise structures and their locations. The act of colouring can also be therapeutic, and can help generate quietness and mindfulness with some of the benefits that meditation imparts on the brain. Anatomy flashcards e.g. Netter's Anatomy Flash Cards and Mosby's Anatomy & Physiology Study and Review Cards can also be helpful.

Online resources: There are lots of great online resources for studying anatomy & physiology including YouTube videos, Apps and online multimedia learning platforms such as:
https://www.getbodysmart.com: Animated text narrations and quizzes to explain the structures and functions of human body systems.
www.kenhub.com: Videos and advanced quizzes
https://www.wiley.com/college/apcentral/anatomydrill/: A collection of study exercises maintained by John Wiley & Sons, Inc.
https://www.innerbody.com/htm/body.html: Features over 1,000 interactive maps along with expert explanations to help students visualise human anatomy and physiology.
https://www.biointeractive.org: Science, stories and data to engage students in exploring the living world.
http://msjensen.cbs.umn.edu/webanatomy/self/: Lots of quizzes of varying lengths and difficulty.
https://www.bartleby.com/107/: "Gray's Anatomy" – the reference book that dates back to 1858 is now online for quick and easy access. Included in the 'virtual' Gray's Anatomy of the Human Body are more than 1,200 colour illustrations and a subject index with 13,000 entries.
https://wordmint.com: Crosswords, word searches etc.

Tips for remembering skeletal muscle vocabulary

Skeletal muscles are named according to a number of criteria, each of which describes the muscle in some way e.g. size, fibre direction, location, shape, number of muscles in a group, number of origins, or the muscle action.

Size: The three muscles that make up the buttocks are named according to their size: gluteus maximus (largest), gluteus medius (medium sized) and gluteus minimus (smallest). In the chest, the pectoral muscles include pectoralis major (largest one) and minor (smallest one) and the smiling muscles in the face are the zygomaticus major and minor – these muscles also reflect their location as they cross the zygomatic bone - cheek bone. Names are often used to indicate muscle length e.g. brevis (short) or longus (long).

Fibre direction: The names of the abdominal muscles indicate the direction of their fibers e.g. rectus abdominis (straight/vertical), internal and external obliques (at an angle) and transverse abdominis (horizontal) muscle fibres.

Location: A muscle's anatomical location or its relationship to a particular bone often determines its name. For example, the frontalis muscle covers the frontal bone of the skull. Other examples are muscles of the arm that include the term brachii (meaning 'of the arm') e.g. biceps brachii, or biceps femoris - one of the hamstrings that crosses the femur at the back of the thigh. Some muscle names indicate their position relative to the midline of the body e.g. lateralis (to the outside or away from the midline) and medialis (towards the midline). The location of a muscle's attachment can also appear in its name. When the name of a muscle is based on the attachments, the origin is usually named first. For example, the sternocleidomastoid muscle of the neck has a dual origin on the sternum (sterno) and clavicle (cleido) and it inserts on the mastoid process of the temporal bone.

Shape: Some muscle names reflect their shape e.g. the deltoid is a large, triangular-shaped muscle that covers the shoulder, and its shape resembles the triangular shaped Greek letter 'delta'. The trapezius muscle is named after its trapezoid shape and the rhomboids muscles that run between the scapulae and the spine are so called after the Greek 'rhombos', which refers to their angled parallelogram shape.

Group number: Some muscle names indicate the number of muscles in a group e.g. the quadriceps is a group of four muscles located on the anterior (front) thigh.

Number of origins: Muscle names can provide information as to how many origins a particular muscle has. For example the biceps brachii and the triceps brachii – the prefix 'bi' indicates that the muscle has two origins and 'tri' indicates three origins.

Action: Flexor muscles are those that decrease the angle at a joint, while extensors do the opposite, increasing the angle at a joint. Abductors move bones away from the midline of the body, while adductors move bones towards the midline.

When we talk about the 'action' of a muscle we are referring to the body action that occurs when that muscle contracts e.g. when the biceps brachii on the upper arm contracts the elbow is flexed and we can bring food to our mouth. **Here are some generalisations about muscle actions – but there are some exceptions!**

Muscles of the anterior arm are generally flexors of the elbow, wrist or fingers e.g. biceps brachii, brachialis, flexor carpi ulnaris, flexor carpi digitorum. The forearm pronators are also located here e.g. pronator teres.

Muscles of the posterior arm are generally extensors of the elbow, wrist and fingers e.g. triceps brachii, extensor carpi radialis, extensor digitorum. Supinator muscles are also located here e.g. brachioradialis (supinator longus), supinator (brevis).

Muscles of the anterior leg are generally extensors & dorsiflexors e.g. quadriceps (extend the knee), dorsiflexors and toe extensors e.g. tibialis anterior, extensor hallucis longus, extensor digitorum longus.

Muscles of the posterior leg are generally flexors & plantarflexors e.g. hamstrings (flex the knee), flexor hallucis longus & flexor digitorum longus (flex the toes), and plantarflexors e.g. gastrocnemius and soleus (calf muscles).

Muscles of the inner thigh are generally adductors e.g. the adductors, gracilis, pectineus.

Muscles of the outer thigh and shoulder are generally abductors e.g. gluteus minimus, gluteus medius and tensor fascia lata abduct the hip, and the deltoid on the shoulder abducts the arm.

Muscles closer to the midline of the trunk (medial) are generally adductors e.g. adductors of arm are supraspinatus, pectoralis major, latissimus dorsi, teres major, and the rhomboids adduct the scapulae.

NOTE: Although fascia - a vast network of connective tissue within the body - has always been with us, it is only recently being appreciated for the invaluable role that it plays in the body's health and functioning. Skeletal muscles are completely surrounded and interwoven with three distinct layers of fascia and these three fascial layers blend together at either end of a muscle and become its tendons. Muscle and fascia are so intertwined that a muscle, collectively with its tendons, is often referred to as a myofascial unit instead of simply a muscle. As well as surrounding muscle fibres and muscles, fascia permeates the whole body and is sometimes described as having the appearance of a spider's web or the membranes of an orange.

Fascia research is offering many new insights into the structure and function of the human body. As this emerging field develops, anyone studying anatomy and physiology would be well advised to investigate the subject of fascia.

You can read more about fascia in the book "KNOW YOUR BODY The Essential Guide to Human Anatomy and Physiology" that accompanies this workbook, and find links to experts on this subject.

ANSWERS

Multiple Choice Tests

Anatomical Terminology

1. c) 2. a) 3. d) 4. b) 5. a) 6. c) 7. a) 8. c) 9. b) 10. c) 11. a) 12. a) 13. c) 14. c)

Body Compounds

1. a) 2. b) 3. c) 4. a) 5. c) 6. c) 7. d) 8. c) 9. c) 10. c) 11. b) 12. d) 13. a) 14. c)

Histology

1. b) 2. a) 3. b) 4. d) 5. c) 6. b) 7. d) 8. b) 9. d) 10. c) 11. b) 12. a) 13. c) 14. d)

Skin

1. c) 2. d) 3. a) 4. b) 5. b) 6. c) 7. b) 8. c) 9. d) 10. a) 11. c) 12. d) 13. c) 14. a)

Hair

1. b) 2. c) 3. a) 4. b) 5. c) 6. c) 7. b) 8. d) 9. d) 10. a) 11. d) 12. b) 13. c) 14. c)

Nails

1. a) 2. c) 3. c) 4. c) 5. a) 6. d) 7. c) 8. c) 9. d) 10. d) 11. c) 12. c) 13. d) 14. b)

Skeletal system

1. b) 2. c) 3. c) 4. c) 5. d) 6. b) 7. b) 8. c) 9. a) 10. d) 11. c) 12. b) 13. a) 14. b)

Muscular System

1. b) 2. c) 3. a) 4. d) 5. a) 6. d) 7. b) 8. d) 9. c) 10. a) 11. d) 12. b) 13. b) 14. c)

Skeletal Muscles

1. a) 2. c) 3. a) 4. b) 5. d) 6. c) 7. a) 8. b) 9. c) 10. d) 11. c) 12. d) 13. c) 14. c)

Cardiovascular System

1. c) 2. d) 3. c) 4. a) 5. c) 6. b) 7. a) 8. a) 9. c) 10. b) 11. c) 12. b) 13. b) 14. a)

Lymphatic System

1. d) 2. d) 3. c) 4. a) 5. b) 6. c) 7. b) 8. a) 9. b) 10. c) 11. b) 12. d) 13. c) 14. d)

Immunity

1. c) 2. c) 3. d) 4. c) 5. a) 6. d) 7. a) 8. a) 9. b) 10. c) 11. a) 12. b) 13. b) 14. c)

Respiratory System

1. b) 2. d) 3. d) 4. c) 5. c) 6. d) 7. c) 8. b) 9. d) 10. c) 11. d) 12. c) 13. c) 14. d)

Digestive System

1. b) 2. c) 3. d) 4. d) 5. b) 6. a) 7. d) 8. b) 9. c) 10. a) 11. c) 12. b) 13. b) 14. c)

Nervous System

1. b) 2. a) 3. b) 4. b) 5. a) 6. a) 7. a) 8. d) 9. b) 10. d) 11. c) 12. c) 13. b) 14. d)

The Senses

1. b) 2. d) 3. b) 4. c) 5. d) 6. c) 7. b) 8. a) 9. b) 10. b) 11. c) 12. c) 13. c) 14. d)

Endocrine System

1. b) 2. a) 3. b) 4. a) 5. c) 6. a) 7. d) 8. a) 9. a) 10. a) 11. c) 12. b) 13. a) 14. d)

Urinary System

1. c) 2. b) 3. c) 4. a) 5. c) 6. a) 7. d) 8. b) 9. d) 10. c) 11. b) 12. a) 13. b) 14. a)

Reproductive System

1. b) 2. c) 3. d) 4. c) 5. b) 6. d) 7. b) 8. a) 9. b) 10. c) 11. b) 12. a) 13. c) 14. b)

Mixed Multiple Choice Test 1

1. c) 2. b) 3. a) 4. c) 5. d) 6. c) 7. a) 8. c) 9. d) 10. d) 11. c) 12. b) 13. a) 14. a) 15. d) 16. a) 17. d) 18. c) 19. c) 20. c) 21. d) 22. c) 23. a) 24. c) 25. c) 26. b) 27. b) 28. d) 29. a) 30. b) 31. a) 32. c) 33. d) 34. b) 35. a) 36. c) 37. a) 38. a) 39. a) 40. d) 41. b) 42. c) 43. d) 44. b) 45. c) 46. a) 47. c) 48. d) 49. a) 50. b)

Mixed Multiple Choice Test 2

1. a) 2. a) 3. c) 4. c) 5. c) 6. d) 7. b) 8. c) 9. c) 10. a) 11. c) 12. c) 13. c) 14. a) 15. b) 16. c) 17. c) 18. a) 19. d) 20. d) 21. a) 22. d) 23. a) 24. c) 25. a) 26. d) 27. a) 28. c) 29. b) 30. d) 31. d) 32. b) 33. a) 34. b) 35. b) 36. d) 37. d) 38. b) 39. d) 40. c) 41. c) 42. b) 43. c) 44. d) 45. d) 46. d) 47. a) 48. a) 49. a) 50. c)

Mixed Multiple Choice Test 3

1. b) 2. c) 3. b) 4. b) 5. b) 6. c) 7. d) 8. a) 9. a) 10. c) 11. a) 12. a) 13. c) 14. b) 15. a) 16. a) 17. c) 18. b) 19. c) 20. b) 21. d) 22. b) 23. c) 24. b) 25. c) 26. d) 27. c) 28. c) 29. c) 30. b) 31. d) 32. b) 33. d) 34. c) 35. a) 36. c) 37. b) 38. c) 39. c) 40. b) 41. c) 42. a) 43. d) 44. b) 45. a) 46. c) 47. c) 48. d) 49. a) 50. d)

Mixed Multiple Choice Test 4

1. b) 2. d) 3. d) 4. c) 5. c) 6. c) 7. b) 8. d) 9. c) 10. a) 11. d) 12. d) 13. d) 14. c) 15. a) 16. b) 17. b) 18. c) 19. a) 20. b) 21. a) 22. c) 23. c) 24. b) 25. a) 26. b) 27. c) 28. b) 29. a) 30. a) 31. d) 32. d) 33. d) 34. b) 35. b) 36. a) 37. a) 38. b) 39. a) 40. d) 41. c) 42. b) 43. c) 44. d) 45. b) 46. c) 47. a) 48. c) 49. b) 50. b)

Anatomical Terminology – Gap Filling

1. Flexion 2. Supination 3. Inversion 4. Plantarflexion 5. Extension 6. Circumduction 7. Abduction 8. Pronation 9. Transverse plane 10. Coronal (frontal) plane 11. Sagittal plane 12. Midline 13. Dorsiflexion 14. Anterior 15. Posterior 16. Medial 17. Proximal 18. Caudal 19. Anatomical position 20. Eversion 21. Elevation 22. Distal 23. Lateral 24. Internal rotation 25. Depression 26. Adduction 27. Cranial 28. External rotation 29. Atom 30. Metabolism 31. Catabolism 32. Anabolism 33. Molecule 34. Differentiation 35. Reproduction 36. Histology 37. Ipsilateral 38. Cranial 39. Contralateral

Histology – Gap Filling

1. Ligaments 2. Cell membrane 3. Nucleolus 4. Cytoplasm 5. Nucleus 6. Tendons 7. Endoplasmic reticulum 8. Nuclear membrane 9. Golgi apparatus 10. Centrioles 11. Interphase 12. Chromatids 13. Neuroglia 14. Telomeres 15. Vacuoles 16. Adipose tissue 17. Fascia 18. Cartilage 19. Centrosome 20. Mitochondria 21. Lysosomes 22. Ribosomes 23. Centrosome

Skin/Scalp Disorders – Gap Filling

1. Acne vulgaris 2. Impetigo 3. Boil 4. Folliculitis 5. Conjunctivitis 6. MRSA 7. Cellulitis 8. Stye 9. Warts, verrucas 10. Herpes simplex 11. Herpes zoster 12. Tinea corporis 13. Tinea pedis 14. Tinea capitis 15. Candida 16. Scabies 17. Lice 18. Vitiligo 19. Albinism 20. Chloasma 21. Ephilides 22. Lentigo 23. Moles 24. Naevae 25. Port wine stain 26. Psoriasis 27. Keloid 28. Seborrheic keratoses 29. Verruca filliformis 30. Allergic reaction 31. Urticaria 32. Eczema/dermatitis 33. Basal cell carcinoma 34. Squamous cell carcinoma 35. Malignant melanoma 36. Anhidrosis 37. Bromhidrosis (osmidrosis) 38. Hyperhidrosis 39. Milaria rubra 40. Keratosis pilaris 41. Seborrheic dermatitis 42. Alopecia areata 43. Striae 44. Crow's feet 45. Broken capillaries 46. UV damage 47. Comedones 48. Milia 49. Pressure sores 50. Scleroderma 51. Systemic lupus erythematosus (SLE) 52. Burns 53. Chilblains

Nail Diseases (Onychosis) - Gap Filling

1. Paronychia 2. Corns 3. Ridges/Furrows 4. Pitting 5. Leukonychia 6. Beau's lines 7. Hangnail 8. Onychorrhexis 9. Stained nails 10. Black nails 11. Blue nails 12. Onycholysis 13. Melanoma 14. Onychatrophia 15.Onychogryphosis 16. Onychocryptosis 17. Koilonychia 18. Onychophagia 19. Dermatitis /Eczema 20. Whitlow 21. Tinea ungium 22. Tinea pedis 23. Psoriasis 24. Onychia 25. Onychauxis

Skeletal System Disorders - Gap Filling

1. Cervical spondylosis 2. Bunions 3. Bursitis 4. Ganglion 5. Prostheses 6. Gout 7. Osteoarthritis 8. Whiplash 9. Osteoporosis 10. Prolapsed (herniated/slipped disc) 11. Rheumatoid arthritis 12. Arthritis 13. Lordosis 14. Ankylosing spondylitis 15. Kyphosis 16. Osteogenesis imperfecta 17. Carpal tunnel syndrome 18. Psoriatic arthritis 19. Paget's disease 20. Osteomalacia 21. Spinal stenosis 22. Fracture 23. Scoliosis

Muscular Disorders - Gap Filling

1. Atrophy 2. Bursitis 3. Cramp 4. Fibromyalgia 5. Adhesive capsulitis 6. Shin splints 7. Atony 8. Adhesions 9. Sprain 10. Strain 11. Tendonitis 12. Tetanus 13. Myositis 14. Microtrauma 15. Tennis elbow 16. Muscle fatigue 17. Rheumatism 18. Repetitive strain injury (RSI) 19. Golfer's elbow 20. Housemaid's knee 21. Lumbago 22. Muscular dystrophy 23. Spasticity 24. Myasthenia gravis 25. Cramp (X2) 26. Plantar fasciitis

Cardiovascular System Disorders - Gap Filling

1. Anaemia 2. Raynaud's disease 3. Cardiac arrhythmia; tachycardia; bradycardia 4. Angina 5. Aneurysm 6. Haematoma 7. Hemorrhoids 8. Arteriosclerosis 9. Atherosclerosis 10. Coronary thrombosis 11. Hypertension 12. Hypotension 13. HIV; AIDS 14. Varicose ulcers 15. Thrombus 16. Palpitations 17. Pulmonary embolism 18. Embolus 19. Phlebitis 20. Hepatitis 21. Haemophilia 22. Thalassaemia 23. Sickle cell anaemia 24. Cholesterol 25. Epistaxis 26. Gangrene 27. Hole in the heart 28. Varicose Veins 29. Intermittent claudication 30. Myocardial infarction 31. Leukaemia 32. Septicaemia

Lymphatic & Immune System Disorders - Gap Filling

1. Hodgkin's lymphoma 2. Non-Hodgkin's lymphoma 3. Allergies 4. HIV; AIDS 5. Glandular fever 6. Tonsillitis 7. Systemic lupus erythematosus (SLE) 8. Lymphoedema 9. Leukaemia

Respiratory System Disorders - Gap Filling

1. Asthma 2. Bronchitis 3. Cancer 4. Common cold 5. Cough 6. Emphysema 7. Hay fever 8. Influenza (flu) 9. Laryngitis 10. Pleurisy 11. Pharyngitis 12. Pneumonia 13. Pulmonary embolism 14. Rhinitis 15. Sinusitis 16. Tonsillitis 17. Tuberculosis 18. Bronchiolitis 19. Cor pulmonale 20. Chronic obstructive pulmonary disease/Chronic obstructive airways disease (COPD/COAD) 21. Cystic fibrosis 22. Hyperventilation 23. Lung cancer 24. Whooping cough (pertussis) 25. Pneumothorax 26. Pulmonary fibrosis 27. Sarcoidosis 28. Severe acute respiratory syndrome (SARS) 29. Snoring 30. Tracheitis

Digestive System Disorders - Gap Filling

1. Anorexia (nervosa) 2. Appendicitis 3. Bulimia (nervosa) 4. Cancer 5. Cirrhosis 6. Jaundice 7. Stress 8. Ageusia / Hypogeusia / Hypergeusia / Ageusia 9. Ulcer 10. Gallstones 11. Gingivitis 12. Heartburn (reflux oesophagitis) 13. Hernia 14. Hiccoughs 15. Indigestion (dyspepsia) 16. Irritable bowel syndrome (IBS) 17. Constipation 18. Coeliac disease 19. Diarrhoea 20. Diabetes mellitus 21. Flatulence 22. Candida 23. Colitis 24. Ulcerative colitis 25. Crohn's disease 26. Diverticulosis 27. Diverticulitis 28. Enteritis 29. Gastritis 30. Cholecystitis 31. Pernicious anaemia 32. Oesophegal ulcer

Nervous System Disorders - Gap Filling

1. Alcohol abuse 2. Bell's palsy 3. Cerebral palsy 4. Epilepsy 5. Drug abuse 6. Headache 7. Migraine 8. Motor neurone disease 9. Multiple sclerosis 10. Depression 11. Clinical depression 12. Bipolar affective disorder 13. Seasonal affective disorder (SAD) 14. Post-natal depression 15. Myalgic encephalomyelitis (ME) 16. Neuritis 17. Neuralgia 18. Parkinson's disease 19. Sciatica 20. Stroke 21. Transient iaschemic attack (TIA) 22. Alzheimer's disease 23. Concussion 24. Meningitis 25. Myasthenia gravis 26. Paralysis 27. Peripheral neuropathy 28. Poliomyelitis 29. Spina bifida

Endocrine System Disorders - Gap Filling

1. Addison's disease 2. Cushing's syndrome/Cushing's disease 3. Diabetes insipidus 4. Diabetes mellitus 5. Goitre 6. Grave's disease 7. Hyperthyroidism 8. Hypothyroidism 9. Insomnia 10. Polycystic ovarian syndrome (PCOS) 11. Myxoedema 12. Stress 13. Acromegaly 14. Gigantism 15. Hyperparathyroidism 16. Hypoparathyroidism

Urinary System Disorders - Gap Filling

1. Benign prostatic hyperplasia (BPH) 2. Cancer 3. Cystitis 4. Diabetes insipidus 5. Gout 6. Urinary incontinence 7. Interstitial cystitis (IC) 8. Kidney stones 9. Nephritis 10. Pyelonephritis 11. Glomerulonephritis 12. Prostatitis 13. Proteinuria 14. Urinary retention 15. Renal failure 16. Urinary tract infections (UTI's)

Reproductive System Disorders - Gap Filling

1. Mastalgia 2. Cysts 3. Galactorrhea 4. Mastitis 5. Chlamydia 6. Ectopic pregnancy 7. Endometriosis 8. Fibroids 9. Menstrual disorders 10. Pre-menstrual syndrome 11. Polycystic ovarian syndrome (PCOS) 12. Prostatitis 13. Vulvovaginal candidiasis (thrush) 14. Hysterectomy

WORD SEARCHES

Word Search 1 (Horizontal & Vertical)

D	G	O	I	T	R	E	D	D	P	B	B	Y	A	A	G	P
J	P	W	E	L	O	G	O	U	N	W	R	D	C	N	R	Q
C	G	I	N	G	I	V	I	T	I	S	O	I	R	E	A	A
H	A	D	L	B	T	F	G	O	U	T	N	A	O	U	L	Z
O	R	E	P	U	G	N	B	X	V	V	C	B	M	R	O	E
L	T	M	O	R	H	U	U	I	E	K	H	E	E	Y	P	M
E	H	E	D	S	A	Y	L	E	C	U	I	T	G	S	E	P
C	R	N	O	I	J	F	I	O	Z	Y	T	E	A	M	C	H
Y	I	T	K	T	E	E	M	H	E	Y	I	S	L	Z	I	Y
S	T	I	J	I	Z	V	I	W	M	R	S	I	Y	T	A	S
T	I	A	K	S	E	E	A	A	A	C	A	N	C	E	R	E
I	S	W	C	I	R	R	H	O	S	I	S	X	K	I	B	M
T	E	N	D	O	M	E	T	R	I	O	S	I	S	H	U	A
I	K	S	C	Y	S	T	I	T	I	S	U	S	N	J	N	S
S	I	F	K	C	Z	G	D	E	R	M	A	T	I	T	I	S
E	P	I	L	E	P	S	Y	P	T	Q	X	M	B	R	O	D
D	I	V	E	R	T	I	C	U	L	I	T	I	S	N	N	O

Word Search 2 (Horizontal & Vertical)

Y	I	G	L	A	H	L	U	M	B	A	G	O	K	L	J	H
A	C	E	I	D	I	N	E	M	M	O	D	K	L	E	R	A
I	K	M	M	K	V	E	A	A	L	I	X	E	N	U	M	E
J	C	I	E	V	E	P	H	S	G	P	T	N	I	K	E	M
L	I	G	N	U	S	H	Y	T	M	L	H	E	N	A	N	O
Y	M	R	I	H	P	R	P	I	E	A	Y	U	S	E	O	R
M	P	A	N	E	K	I	O	T	N	R	P	R	O	M	R	R
P	E	I	G	R	E	T	T	I	O	Y	E	I	M	I	R	H
H	T	N	I	N	Q	I	H	S	P	N	R	T	N	A	H	O
O	I	E	T	I	G	S	Y	C	A	G	T	I	I	N	A	I
M	G	C	I	A	J	N	R	Y	U	I	E	S	A	X	G	D
A	O	J	S	Y	V	F	O	G	S	T	N	A	G	M	I	S
V	I	J	A	U	N	D	I	C	E	I	S	C	J	U	A	G
U	A	L	H	P	B	U	D	F	P	S	I	N	B	M	L	A
N	E	U	R	A	L	G	I	A	A	N	O	E	V	P	W	W
G	P	L	V	N	A	U	S	E	A	B	N	N	L	S	Z	P
K	O	V	H	M	Q	G	M	X	I	M	E	A	S	L	E	S

Word Search 3 (Horizontal & Vertical)

I	P	X	O	Y	P	L	E	U	R	I	S	Y	I	K	Z	R
O	T	I	T	I	S	S	W	L	C	E	K	N	M	F	W	O
O	S	H	I	N	G	L	E	S	C	A	B	I	E	S	O	S
S	N	E	U	R	A	L	G	I	A	P	C	K	H	I	S	A
T	S	K	R	I	C	K	E	T	S	O	L	X	B	M	T	C
E	J	J	J	W	I	S	C	O	L	I	O	S	I	S	E	E
O	H	L	O	S	T	E	O	P	O	R	O	S	I	S	O	A
M	L	O	R	D	O	S	I	S	E	P	S	I	S	C	A	K
A	I	P	S	O	R	I	A	S	I	S	O	E	H	R	R	Y
L	P	S	C	I	A	T	I	C	A	S	E	L	U	H	T	P
A	H	I	Y	V	L	N	B	F	F	L	D	G	B	I	H	H
C	P	S	I	N	U	S	I	T	I	S	E	E	Q	N	R	O
I	C	K	R	I	N	G	W	O	R	M	M	F	D	I	I	S
A	S	E	P	T	I	C	A	E	M	I	A	K	N	T	T	I
L	Z	H	P	R	O	S	T	A	T	I	T	I	S	I	I	S
P	D	P	N	E	U	M	O	N	I	A	D	P	D	S	S	B
J	G	N	P	A	L	P	I	T	A	T	I	O	N	S	B	C

Word Search 4 (Horizontal & Vertical)

F	T	H	R	O	M	B	O	S	I	S	T	E	N	D	O	N
U	R	T	I	C	A	R	I	A	N	V	N	H	E	N	D	P
X	S	F	C	X	Y	B	S	T	R	E	S	S	W	U	G	Y
K	P	K	H	O	U	S	P	O	N	D	Y	L	O	S	I	S
S	R	V	L	A	L	G	S	R	F	A	S	C	I	A	F	V
D	A	A	O	Z	C	S	Y	P	H	I	L	L	I	S	L	A
C	I	K	A	V	E	R	R	U	C	A	V	I	P	I	E	R
I	N	V	S	N	R	G	T	H	R	U	S	H	C	T	P	I
T	Y	E	M	O	U	U	I	T	I	N	N	I	T	U	S	C
Y	X	R	A	P	W	L	I	G	A	M	E	N	T	O	M	O
V	I	T	I	L	I	G	O	B	N	S	T	R	A	I	N	S
O	T	I	T	U	B	E	R	C	U	L	O	S	I	S	X	E
G	R	G	S	P	O	N	D	Y	L	I	T	I	S	H	O	Z
N	B	O	X	D	U	E	S	Y	P	H	I	L	I	S	K	F
N	G	B	Q	X	Q	W	A	R	T	T	R	V	R	O	E	A
J	T	O	N	S	I	L	L	I	T	I	S	V	Y	Z	D	X
O	S	T	R	O	K	E	F	T	E	T	A	N	U	S	H	C

CROSSWORDS

Anatomical Terminology

Across:

3. Chest **6.** Pathology **8.** Risk factor **11.** Zoonotic **12.** Red flag **13.** Pelvis **14.** Syndrome

Down:

1. Contralateral **2.** Idiopathic **4.** Epigastric **5.** Sagittal **7.** Transverse **9.** Chronic **10.** Seven

Body Compounds

Across:

2. Retinol **4.** Pernicious **10.** Carbohydrates **11.** Iodine **13.** Seven **14.** Calcium

Down:

1. Sodium **3.** Sunlight **5.** Iron **6.** Hydrogen **7.** Fats **8.** Oxygen **9.** Water **12.** Organic

Histology

Across:

6. Telomeres **8.** Histology **9.** Nondisjunction **11.** Osmosis **12.** Telophase **13.** Phagocytosis **14.** Metaphase

Down:

1. Genes **2.** Mitochondria **3.** Lysosomes **4.** Centrioles **5.** Translocation **7.** Fascia **10.** Ribosomes

Skin

Across:

2. Apocrine **4.** Hyperhidrosis **7.** Sebum **8.** Dermis **10.** Basal **12.** Keratin **13.** Bromhidrosis **14.** Constrict

Down:

1. Macrophages **3.** Cholecalciferol **5.** Dehydrated **6.** Adipose **9.** Melanin **11.** Eccrine

Skin Disorders

Across:

3. Hemangiomas **6.** Melanoma **10.** Lupus **12.** Varicella **13.** Shingles **14.** Psoriasis

Down:

1. Scleroderma **2.** MRSA **4.** Warts **5.** Xanthomas **7.** Eczema **8.** Impetigo **9.** Vitiligo **11.** Melasma

Hair

Across:

2. Lanugo **7.** Medulla **9.** Melanin **10.** Keratin **12.** Cortex **13.** Catagen **14.** Hirsutism

Down

1. Alopecia **3.** Anagen **4.** Headlice **5.** Matrix **6.** Telogen **8.** Cuticle **11.** Shaft

Nails

Across:

2. Koilonychia **6.** Onychosis **7.** Paronychium **9.** Cyanosis **11.** Onychorrhexis **12.** Eponychium **13.** Psoriasis **14.** Onychophagia

Down:

1. Ringworm **3.** Whitlow **4.** Hyponychium **5.** Onychocryptosis **8.** Lunula **10.** Keratin

Skeletal System

Across:

2. Comminuted **4.** Arthritis **7.** Synovial **9.** Haversian **12.** Gout **13.** Compact **14.** Periosteum **15.** Cheekbones

Down:

1. Arthrology **3.** Osteomalacia **5.** Osteoclasts **6.** Mandible **8.** Ossification **10.** Spongy **11.** Osteoporosis

Muscular System

Across:

3. Fibromyalgia **5.** Abduction **7.** Acetylcholine **8.** Neuromuscular **10.** Botox **11.** Belly **12.** Myosin **13.** Isometric **14.** Smooth

Down:

1. Myofibril **2.** Cardiac **4.** Extensibility **6.** Myositis **9.** Epimysium

Skeletal Muscles

Across:

2. Achilles **9.** Masseter **11.** Soleus **12.** Hamstrings **13.** Deltoid **14.** Quadriceps

Down:

1. Rhomboids **3.** Zygomatic **4.** Procerus **5.** Platysma **6.** Piriformis **7.** Iliopsoas **8.** Extensors **10.** Trapezius

Cardiovascular System

Across:

5. Phagocytes **10.** Thymus **11.** Eosinophils **12.** Landsteiner **13.** Systolic **14.** Anaemia

Down:

1. Carotid **2.** Haemoglobin **3.** Erythropoietin **4.** Pacemaker **6.** Atherosclerosis **7.** Platelets **8.** Tricuspid **9.** Sepsis

Lymphatic System & Immunity

Across:

3. Glandular **7.** Axillary **10.** Adaptive **11.** Antigens **12.** Spleen **13.** Supratrochlear **14.** Lymphomas

Down:

1. Innate **2.** Antibodies **4.** Lacteals **5.** HIV **6.** Lymphedema **8.** Popliteal **9.** Anaphylaxis

Respiratory System

Across:

2. Larynx **6.** Rhinitis **9.** Sinuses **10.** Phrenic **12.** Medulla **13.** Pleura **14.** Epiglottis

Down:

1. Pertussis **3.** Histamine **4.** Pneumonia **5.** Tuberculosis **7.** Coronavirus **8.** Asthma **11.** Alveoli

Digestive System

Across:

3. Duodenum **5.** Diverticulosis **8.** Vagus **10.** Lipase **11.** Cardiac **13.** Enteritis **14.** Bolus

Down:

1. Peritoneum **2.** Incisors **4.** Hepatitis **6.** Hydrochloric **7.** Ingestion **9.** Pyloric **12.** Amylase

Nervous System

Across:

2. Olfactory **5.** Autonomic **8.** Sympathetic **10.** Endorphins **12.** Cerebellum **13.** Somatic **14.** Meninges

Down:

1. Dopamine **3.** Neurotransmitters **4.** Trigeminal **6.** Myelin **7.** Hippocampus **9.** Myotome **11.** Neuralgia

The Senses

Across:

2. Temporal **8.** Umami **9.** Thermoreceptors **11.**Vestibulocochlear **12.** Ossicles **13.** Macula **14.** Cataract

Down:

1. Glaucoma **3.** Nociceptors **4.** Touch **5.** Cornea **6.** Eustachian **7.** Anosmia **10.** Five

Endocrine System

Across:

2. Cortisol **5.** Prolactin **10.** Adrenaline **11.** Mellitus **12.** Pancreas **13** Oxytocin **14.** Aldosterone

Down:

1. Homeostasis **3.** Thyroid **4.** Vasopressin **6.** Insipidus **7.** Goitre **8.** Parathormone **9.** Pineal

Urinary System

Across:

2. Urea **4.** Dialysis **9.** Ureters **11.** Nephroblastoma **12.** Pelvis **13.** Cystitis **14.** Incontinence

Down:

1. Nephrons **3.** Micturition **5.** Glomerulus **6.** Prostatitis **7.** Enuresis **8.** Erythropoietin **10.** Proteinuria

Reproductive System

Across:

3. Gametes **8.** Fibroids **10.** Preeclampsia **11.** Endometrium **13.** Zygote **14.** Placenta **15.** Amenorrhea **16.** Cervix

Down:

1. Hymen **2.** Mastitis **4.** Hysterectomy **5.** Vagina **6.** Flagellum **7.** Epididymis **9.** Ectopic **12.** Dysmenorrhea

Acknowledgements

I am forever grateful to the many students I have taught in my role as a complementary therapy lecturer over the past 20 years. Your eagerness, curiosity, interest and thought provoking questions kept me learning too, and helped me become a better teacher.

During my years of private practice, the trust of my clients gave me the confidence and encouragement to seek a better understanding of the human body and health issues in general - thank you for trusting me to be your therapist.

To Gabriel Mojay – thank you for the opportunity to teach anatomy and physiology to your students and for writing the foreword to the accompanying book "KNOW YOUR BODY – The Essential Guide to Human Anatomy and Physiology".
To my friends and colleagues who read the accompanying book and made helpful suggestions and comments: Dr. Michael Fenster, MD, Dr. Kaur T. Birinder, MB, BS, MRCOG, Lesley Hart, Emma Holly, Emma Charlton - thank you all for your invaluable feedback, advice and support.

To Grainne Aldred – for designing and creating the eye-catching book covers: http://www.grainnealdred.co.uk

To Jo Harrison – for formatting: https://www.ebook-formatting.co.uk

To Angela Pruden – for proofreading: https://www.facebook.com/Angela-Pruden- Proofreading-103981077759046

To my husband Patrick, thank you for your support and helpful advice.

About the Author

Following a career in education, Mary Dalgleish qualified as an aromatherapist at the Tisserand Institute in London in 1999 and later worked there as an anatomy and physiology lecturer and student mentor. Her aromatherapy studies sparked an interest in natural health, and since then she has completed a variety of complementary therapy training courses including reflexology and various forms of face and body massage.

Mary has worked as a lecturer in the complementary therapy departments of several London adult education colleges including Kingston, Merton, Sutton and Morley Colleges. She has also lectured at several private colleges, including Maureen Burgess AOR Accredited School of Reflexology, Westminster School of Yoga, The School of Natural Therapies, and more recently at the Institute of Traditional Herbal Medicine & Aromatherapy (ITHMA) based at Regent's University in London. Mary also taught a range of continuing professional development courses for qualified therapists.

Mary was awarded a 'Certificate of Excellence' for teaching anatomy & physiology at Merton Adult College in May 2005 and again in May 2008 (nominated by her students). She was also awarded a 'Commended Tutor of the Year' Certificate at the Federation of Holistic Therapists 'Excellence in Education Awards' in 2012.

Mary has co-authored two books to date "Indian Head Massage – the Essential Guide" & "Ear Candling – the Essential Guide" (both originally published by Hodder Arnold as part of a series). Mary is a Senior Fellow and former Vice-President of the Federation of Holistic Therapists, and holds a Master's Degree in Education. During her career, she regularly contributed articles to various health and wellbeing magazines and ran a private clinical practice alongside her teaching and writing. Mary has now retired from clinical practice and lives in the Cotswolds where she enjoys an active lifestyle. She also volunteers as a complementary therapist with a local cancer charity.

As well as this workbook, Mary has written an accompanying book entitled 'KNOW YOUR BODY – The Essential Guide to Human Anatomy and Physiology', which is essential reading for anyone interested in the human body and how it works. Rated by Dr. Michael Fenster, MD FACC, FSCA&I, PEMBA as "A deep and expansive addition to any library; it is an edition no serious student of human health should be without!"

Personal note from the author:
Thank you for reaching the end of this workbook. I hope you have enjoyed it and found it useful. Please spread the word about "KNOW YOUR BODY The Essential Guide to Human Anatomy and Physiology" and feel free to get in touch with any queries or comments relating to the book or workbook.

You can contact me via:
https://anatomyandphysiology.co.uk
Mary Dalgleish

Made in United States
North Haven, CT
10 November 2026

22088259R00098